What the experts are saying about *High Tech with Low Risk*:

"The Northwest is a hotbed of high tech entrepreneurship. It will become the Silicon Valley of the information age, and Trudel's techniques and advice will help make this happen."

Douglas O. Reudink, PhD.
Director
Electronics Technologies
Boeing Aerospace & Electronics

"John's business development insights and methods serve as a guiding light in today's chaotic business environment."

John Stoltz
President
Advanced Systems Concepts, Inc.

"The information economy offers unprecedented opportunity to rural Oregon, and Oregonians **are** willing to venture into the new economy. There are new high tech businesses in Bandon, Hood River, Redmond, Glide, and many other communities struggling to emerge from a natural resource based economy that can no longer deliver the prosperity of the past. The good advice in this timely book can help improve their chances of success."

Anne S. Berblinger
Representative for Oregon
Economic Development Administration
U.S. Department of Commerce

"John Trudel's *High Tech with Low Risk*, written from many years of successful experience, lucidly presents the fundamentals for success in high technology business. Not only are the elements fully described but they are clarified by many examples of strategies resulting in success or failure. To fully understand high tech business, one must read this book."

Dr. Larry Blake
Former President
Oregon Institute of Technology

(continued on back cover)

First edition published 1990.

Published by Regional Services Institute, Eastern Oregon State College.

Printed in the United States of America.

ISBN 0-9626772-2-1

HIGH TECH

with

LOW RISK

(Venturing Safely Into the 90s)

by John D. Trudel

Table of contents

High Tech ...

To Patricia the candid
and, memorially,
to Howard the caring
and for
all the Skunks and Innovators.

Foreword

This book is essential reading for anyone who wishes a better understanding of high technology business. The writer is to be complimented for his concise and realistic treatment of a complex topic. He certainly knows high tech, and has diagnosed its problems and solutions well.

High Tech with Low Risk fills a major void. There are very few competitive books in this field, and surely none as complete and entertaining as this one. The advice sections are outstanding, such as on skunk works and on managing small groups. The book is very strong in operations, i.e., "how to get from here to there."

The book is rich with examples, and it presents a good mix of practice and theory. It illustrates the results of those who have succeeded in the toughest and most uncertain markets in the world, and describes how they won. It also describes typical failures in a positive and informative manner. The quotes are appropriate, and add considerably to the text.

I am confident that you will find this book to be rewarding and educational. The writer's style is effective, it's fun to read, and it comes to an end too soon.

> Bob Davis
> Professor
> Graduate School of Business
> Stanford University
> July 18, 1990

High Tech ...

Preface

The author wishes to acknowledge the support from the Regional Services Institute of Eastern Oregon State College that made this book possible. Personal thanks are extended to "Terry and the Pirates," especially Ms. Annette Johnson, Director of the University Center for Rural Oregon Diversification and a former worker in high-technology industry.

Mr. Terry Edvalson, Director of this Institute, is doing many innovative and entrepreneurial things to further the healthy industrial development of rural Oregon. As we move into the 90s, efforts like his will allow Oregon to develop its latent potential as a healthy environment for knowledge-based companies.

Today, with air transport, FAXs, powerful and inexpensive computers, and global communications, it is no longer necessary for industrial companies to cluster around the smokestacks of dense urban environments. Already rural Oregon has leading-edge companies in areas like high-technology business development, satellite-scan technology, membrane technology, precision navigation and simulation, parallel computing, all types of software, display technology, specialized semiconductors, communications, and many others. As yet most of these companies are small and fragile, but with the help of Terry, Annette, and people like them, some of these will become well known employers by the end of the decade.

Who should read this book?

The fundamental purpose of *High Tech with Low Risk* is to help managers and professionals enjoy more financial success in their technology-enabled business endeavors. We expect that financial backers and stakeholders in such ventures also will find this book to be of value. Students in college business and engineering schools may use it to understand better the subtle factors that govern success in today's competitive, turbulent, and global marketplace. It is perhaps ironic that the closer one works with hard science and technology, the more success seems to depend on combining technical knowledge with softer skills such as vision, intuition, perception, and caring. This book can help you compete more effectively in the markets of the 90s if you are willing to invest and to empower small talented teams.

The author

John D. Trudel is founder and director of The Trudel Group, a high-technology business-development consulting firm. Mr. Trudel enjoyed a successful early career as a technologist for Collins Radio Company, Sanders Associates, E-Systems, and others. He has been a principal in four successful high-tech new ventures, and enjoyed a long career at Tektronix where he played key roles in business venturing and new product development for several company divisions. He introduced the company's first telecom test sets. These later grew to become Tek's Redmond division, and during Mr. Trudel's tenure these products had the highest growth rate and profit margin of any product line in the company. He moved from this job to one in oscilloscopes, and defined and introduced the company's first digital oscilloscope. Mr. Trudel was responsible for planning the most successful product line Tektronix has had for two decades, the 2400 series. Mr. Trudel played key roles in a long string of product successes that have, so far, generated close to $1 billion. Mr. Trudel's work in oscilloscopes represents one of the few cases where a Western firm has profitably recovered market share from the Japanese. Not coincidentally, he has spent much time in Japan hosted by Sony-Tektronix, and has given lectures on Japanese productivity and management technique.

High Tech ...

Introduction

High technology has been, cyclically, either the holy grail or the whipping boy of business pundits. A few ventures enjoy incredible success, but a considerable percentage fail. Management theory holds that proper process can lead to success but, even if true, high technology seems immune to this. Some companies exercise "good business practices" and fail dismally in high tech, yet, paradoxically, some succeed, and still others somehow prosper without using them. There are many cases where high-tech companies have done themselves severe damage by adopting management techniques that worked well for other industries. Successful companies usually prefer to repeat, often to institutionalize, what has worked for them, but high-tech markets can change systemically, and this logical habit may lead to tragedy. Creative destruction runs rampant in the industry, and the strongest companies can be defeated by a weaker opponent offering a better solution. The industry is replete with stories of students who got poor grades for their class projects, but later became millionaires from starting businesses based on these same ideas.

> "Even when it is based on meticulous analysis, endowed with clear focus, and conscientiously managed, knowledge-based innovation still suffers from unique risk and, worse, an innate unpredictability."
>
> — Dr. Peter F. Drucker
> Management Process Expert

As we move into the last decade of the 20th century, success in high technology is more elusive than ever. Large U.S.

corporations have traditionally been based on the exploitation of mass production and mass marketing. With this mindset, product improvement was a strategy of last resort — that is, until the competition forced us to — since upgrading raised costs and reduced profits. The Fortune 500 has never done well at innovation, their record is getting steadily worse, and most are now lucky to invent a single major *new* (non replacement) product a decade. Loss of jobs in large U.S. companies has been steady since 1970, and the replacements are mostly new companies and the foreign "screwdriver plants." These are hollow shells that produce only low-level jobs, but reduce "trade friction." The name comes from how the local value is added. Tested and calibrated boards are imported, screwed into a case bearing a "made in U.S.A." label, and shipped. In 1989, foreign-owned plants in the U.S. created more jobs (and paid their assembly workers better) than did domestic factories.[1]

Myths and misconceptions are common:

- To succeed in high technology requires taking extreme risk.

- The large corporations have a "lock" on technology business.

- It reduces risk to collect information until you have consensus or convince management that the action desired is prudent.

- If you build the "better mousetrap" the world will beat a path to your door.

---

[1] Source: **Business Week**, "Innovation 1990" special issue, June 15, 1990.

- We lose to the Japanese because they have better technology.

All these assumptions are wrong. The Japanese succeed, but they are very conservative. The rapidly changing niche markets of the 1990s favor small, quick, companies, so anyone can win with the right product. One enabled mind with the right information can make better decisions than any committee, because the right decision too late is the wrong decision. The "better mousetrap" assumption is dangerous, because it is a security blanket for misguided inventors. New markets need to be created, and it takes much clever work to do so. Many Japanese managers I have spoken with think the main weakness of the U.S. is poor product targeting. There is convincing data that our technology-enabled product failures are most commonly caused by poor market fit. We do not target our customers' needs very well. We have good ideas and good devices, but we don't transform these into products as crisply as our competitors do. Our technology base is eroding, but our failures are usually caused by missing the market, not by weak technology.

One bright spot has been the phenomenal growth of jobs in small companies, but this is now choking off for lack of funding. Capital has all but dried up for electronics-based new ventures, and takeovers at distress-sale prices are on the rise. The present generation of U.S. ven-

> "If there is a distaste for foreign capital investment, the solution is to get our own house in order."
>
> — *Michael H. Armacost*
> *U.S. Ambassador to Japan*
> *Speech in Portland, Oregon*
> *April 30, 1990*

ture capitalists is doing poorly, and more and more technology investment is foreign. Foreign investors frequently take part of their return in technology access, so unless someone can develop better mechanisms to balance trade we can expect more screwdriver plants.

> "Every promising small company is going to be forced to seek capital overseas."
>
> — *James H. Clark*
> *Chairman*
> *Silicon Graphics, Inc.*

The dreams are dying, replaced by grimmer corporate realities. One used to hear stories like Mitch Kapor's securing the financing for Lotus by writing his business plan on a napkin over dinner, or Steve Jobs and The Woz changing the world. These people were real, human, and you could talk with them, like them, respect them, and think "I can do that too if I just used my brain." I never asked Mitch if the napkin story was true, but, if not, it should have been. I was a spectator when Steve stood tall against the giant screen with the woman throwing the hammer and the audience cheered. What I remember most is the guarded rooms with rows of MACs, and walking by at a quiet time to find a bored guard tentatively rolling a mouse. I didn't even like the machine much, but I could feel the world changing. Time has passed, and these people have moved on. There are still successes — Apollo, Sun, Mentor, Sequent, and many more — but it feels different now. There's less excitement, and more skepticism. The survival rate of high-tech products makes the voyage of the Titanic

> "Death is nature's way of suggesting that you reevaluate your strategy."
>
> — *Jon E. Connell*
> *Senior Vice President*
> *Harris Corporation*

seem successful, but it doesn't need to be that way. Opportunity is everywhere, but people keep making mistakes.

This book is a compendium of fundamentals for success in high-technology business, and a guide for *consistent* success at new products. It mostly discusses working in areas where the technology or the market or *both* are new. See Figure 1. Most products are simple replacements or updates to existing products, for example the annual model changes in the auto industry. When you attempt extensional products or markets, uncertainty is higher. Examples of extensional products include the

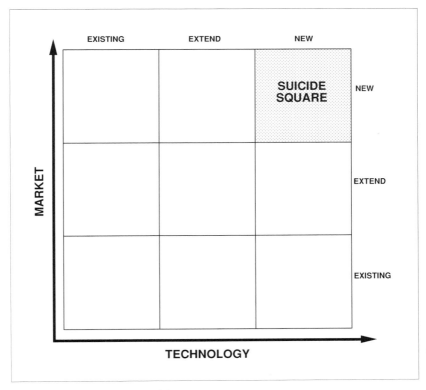

Figure 1. *The playing field.*

change from analog to digital oscilloscopes (extensional technology) or distribution and product changes to reach new customers (extensional markets), or selling cheap seats on jet airliners (extensional products and markets). Uncertainty is higher when either the product or the market is new, and highest when both are. This land of uncertainty is exactly where the most lucrative high-tech opportunities occur.

We in the trade call the area where both the market and the technology are new "the suicide square." Here is where the legends are built, but for each winner there are several losers. If you work in this square, you have to be smart, quick, and meticulous. Your products have to be "right" the first time. You don't get second chances to create first impressions, and if you appear to be succeeding you can be certain that there will be a wave of large and competent competitors hard on your heels. This text will show you how to succeed consistently in new areas, and how to work there with low economic risk. If you can do that well, normal business technique should let you do the rest.

The winners in new high-growth areas can gain dominant positions in large markets and leapfrog established companies. Unfortunately, the losers usually fail totally and consequently write off tens or hundreds of millions of dollars. Often they disappear as companies, and are forgotten. The PC, the minicomputer, desktop publishing, the integrated circuit, and the application-specific integrated circuit (ASIC) were all new technologies that allowed new markets to be created. The companies that won these markets now share multi-billion dollar opportunities.

Is it possible to pursue high-tech opportunity with low risk? Is it possible to ensure that your new products will be consistently successful? It is not only possible, but already many companies are doing it. I will show you actual examples of companies with new product success rates of over 90%. Some are self-funding growth well into the double digits, and some are able to finance and manage compounded growth rates exceeding 100%. If you will take the effort to do the following things well, you can make most of your new products successful, even in the "suicide square":

1. Use the right organizational form. There is good data to show that large organizations take significantly longer to get their products to market, and the delay is attributed to bureaucracy. Smart, flat, fast moving teams are your best hope for success in the information age.

2. Apply ample and timely funding, but with proper controls. Investing too much money can be as bad as not investing enough, and there is no way to recover time lost due to funding delays.

3. Seek out and empower the right diversity of talent. The critical strategic resource today is enabled minds. As a minimum you need a creative leader, a technologist, and a marketer. You will win or lose based on their knowledge, skill, and intuition, and if this is lacking all the management processes in the world can't save you.

4. Develop a portfolio of ventures. Mix your investments between product replacements, extensions, and new. Consistently invest in the future. You need **both** continuous improvement (the Japanese call it *Kaisen*) and new things to be successful.

5. Recognize the difference between creators and implementers. Successful innovation requires both invention and implementation. You need both breeds of people, and you should employ them synergistically.

6. Become globally competitive at manufacturing. If you can't build your product competitively, you can't convert your innovation to profit.

7. Sell the product! The only worthwhile source of revenue for a company is sales. No matter how good your product is, you must generate orders if you hope to make any money.

8. Pay attention to why certain factions are winning consistently. What are we missing about the Japanese? Why is Cypress posting high margins and growing at over 100% per year when most semiconductor firms are declining? Why is Sun so prosperous when DEC and IBM are having problems? Why does Hewlett Packard succeed when its former rival Tektronix is fading?

9. Learn the difference between ventures and adventures. Ventures are nimble, knowledge-based business endeavors in high uncertainty market or technology areas. Adventures are wild gambles

where small teams are jammed against large, resource-rich organizations (either your own, or your competitors). You're deluding yourself if you expect your new ventures to compete *directly* with mature businesses. New things are small and fragile at first. Products nearing the end of their life cycles often have very good financial performance, while new things may be marginal at the outset.

10. Locate and study several good role models. If a company is consistently succeeding in your industry, learn why. Companies like Hewlett Packard, Cypress Semiconductor, Sun, Motorola, and the Japanese are good to learn from.

11. Recognize and understand the most common reasons that U.S. companies fail at new ventures, and avoid them. There are many reasons we fail, but some patterns emerge. We are rarely as diligent at matching our technologies to needs in ways that offer business advantage as some of our competitors. Our investment horizon is short term, and we tend to speculate rather than create wealth. We tend to invent well, but implement poorly. Recall the cautionary tale of the tortoise and the hare. We have the best technology and the best technologists in the world, but I fear that our targeting and follow-through is not very good.

12. Develop your own insightful vision of the future, and do not force it to remain constant. The only thing that remains constant is change. Test your vision continually, and use it to drive your planning

and investments. Continually change and improve.

Caution and disclaimer

This monograph is the anecdotal opinion of one successful professional. I don't know how to write a cookbook process to assure success at high-technology business, and doubt such a thing is possible. You don't want prescriptive answers because they don't apply. Also, I see little value in tediously repeating limitations, cautionary notes, or citing lists of references and examples. High-tech business development is an *art*, so in the end everything comes to knowledge, skill, and intuition. That's what makes it fun, and so lucrative for those who win.

The fundamentals apply, but the results depend on what you do and the talent you can attract. People like to believe skills can be learned, but after a certain level this is not true. Just as few mortals can ever become concert musicians or conductors, so world class skills in the art of high-technology business are elusive. This book will help you establish an orchestra, but the sound will depend on whom you select, how well they perform together, the score chosen, and how well your leader conducts. *People are the critical resource for knowledge-based business, and you ignore this at your peril.*

I've cited a few books and articles as references. Many good sources cannot be listed, since much of the best work is never published. It's common in books such as this to use examples from the author's successes, but I think educational examples more often come from expensive mistakes. I can assure the reader that all examples, positive and negative, stem from actual events. Negative examples are disguised both to protect the guilty — some of whom have suffered enough — and to protect myself from their attorneys.

My desire is not to criticize, but to describe and show how the United States can become more competitive in the technology-based markets of the 1990s. My hope is that some U.S. technology companies can regain the energy we demonstrated in the 70s, and that some readers of this book will go out and change the world for the better.

High Tech …

The right form

A useful model for a high-technology venture is the "skunk works," a uniquely American, astonishingly effective, but rarely used (except during wartime or by venture capitalists) organizational form. The keys are to have a **small** team, that is **talented** in key areas, and **empowered** to move **rapidly** to a **difficult** goal. The first documented skunk works was during the Civil War. After over two decades of barren Federal programs, the Norfolk Navy Yard in 1861 hired a Swedish engineer, a consultant if you will, to design the USS Monitor. The prototype was launched in 126 days, and fought its famous battle a few weeks later. The next known skunk works was opened by Thomas Alva Edison in 1876 to create "inventions to order."

The skunk works

A skunk works requires that at least three roles are present: the sponsor/leader, the technical creator, and the market Guru. Only a trusted and powerful corporate sponsor can commit resources without constant justification. Only talented engineers can create new forms from emerging technology. Only marketing experts can detect latent opportunity before it exists. These roles can be combined in one Renaissance genius, such as Edison, but more likely they are separate people. It is important to keep financial "burn rate" low without skimping on the quality of the people assigned, so *in the early phases it is often better to avoid assigning senior personnel on a full time basis.* You should balance effectiveness and training. You need seasoned people on the team to ensure the correct decisions are made at the "time of creation," but you also should permit learning and skill transfer within the team to take place. Some junior people on today's teams will be tomorrow's leaders or consultants, and all on the team can learn from each other.

> **Note:** Many U.S. managers seem to think that only full-time staff can be truly dedicated to a project. This traditional approach is, I think, incorrect. I think it is better — and much cheaper — to phase appropriate talent into and out of the team as it is needed. If this is done properly, it can lower risk and speed results. First-rate companies make every effort to put the best talent available on their teams, and they are quite flexible and creative about how they do this. Later in this book we will discuss the advantages of shifting the mix of the team over the life of the project in more detail.

Be aware that the members of your team likely have fundamentally different agendas. At the core, engineers are problem solvers. They apply technology to solve problems. The more complex and challenging the technical problem the better they like it, and I've had many tell me they value peer recognition over business success. Marketers seek to understand customer needs, and bend technology to meeting them in ways that offer profit and competitive advantage. The marketer seeks opportunity and wants business success. They want differentiated products, but usually resist using new technology in areas that don't provide major advantage. I have seen projects where team conflict was high over issues such as power supplies, handles, and knobs. To the engineer these offer opportunity for creativity. To the marketer these may be distractions that create risk and delay, but provide little customer value.

Comment: This paragraph drew howls of protest from engineers who cited past disasters at their companies. "Remember the debacle when the power supply on the (name deleted) product line proved unreliable?" Indeed I did, and that is why I wrote these words. These types of disasters cost their companies fortunes when products fail (or catch fire or fall apart) in customer's hands all over the world. Generally the part that failed was a new design, *though it didn't need to be.*

Having a new design was better, but it wasn't an important part of the product. Most companies have these stories, and if you dig deeply you usually find that the designer assigned was inexperienced *because it was not important.* The designer started over from scratch, and most likely he didn't fully understand and appreciate the existing design. I agree that functionality, quality, and reliability are needed. All the more reason, I think, to stick to (or, better, adapt and improve) existing designs when practical. Evolution is much safer than revolution. It's fun to design new things, but it may be better business to remember, "If it ain't broke, don't fix it." Many "new" products or ideas actually evolve from existing ones, and the best are spurred on by unfulfilled customer needs.

The leader should arbitrate conflicts in ways that maximize the probability of business success. To do this they must have intimate knowledge of value and technical risk, and the ability to make engineers *want* to focus on the problems that offer advantage. It is educational to compare design reuse in Japanese and American product lines. We frequently err on the side of technical arrogance, and like to start with a clean sheet each time. The Japanese are masters of adaptation, and often (not always — often) improve or reuse existing designs. The leader is responsible for applying his limited resources where the most value is added, and for getting to market rapidly with the right product.

The team

How can you be sure if you have put together the right type of team? A check list may be helpful:

☐ Are experienced, top quality, engineers on the team? Do these people have a good track record of technical creation?

☐ Are experienced, top quality, technical marketers (not salespeople) on the team? Do these people have a good track record of market creation?

☐ Do you have a seasoned and creative team member responsible for manufacturing issues?

☐ Is there a trusted leader/sponsor on the team? Is this his or her primary or sole responsibility? (You can have the leader fill in for others on the team, but it's dangerous to make him serve a separate organization.)

☐ Is funding adequate and, more important, stable and patient?

☐ Are there meaningful rewards for individual performance?

☐ Can all team members satisfy their personal and professional agendas?

Unless every box gets an honest check, you are making life unnecessarily difficult for yourself, your team, and your company. The leader excepted, it doesn't matter where your talent comes from, but it is crucial that all roles are in place at the right time. Once you have the team together, you should not disassemble it. You should let it work together, gain knowledge, and produce results.

Comment: In the U.S. we often demonstrate what I call the Hollywood or "Rocky" mentality. We like to believe that uneducated, untrained, and inexperienced teams with a "good engineer" and a "better mousetrap" can beat world-class competition. We hope for a magic talisman that will allow a down-and-out manager to assemble a random group of "winos" and beat the best companies in the world.

Sometimes improbable things happen, but technology-enabled business is the fastest, nastiest, and most fun game on the planet. To be a first rate — or even a second rate — company in high tech, you must succeed consistently. Luck helps, but you need talent, investment, and commitment to win consistently.

I am all in favor of training, since this is one of the two ways for consultants to make good incomes. I do not, however, advise my clients to set their rookie teams against world-class competition in the "suicide square." One time in a hundred such a team might win, but losing the other 99 times will waste your scarce resource, grind up your promising young talent, and erode your corporate reputation. There are better ways to season your troops.

You need your best talent in place early, since the decisions made at the time of creation are critical. Most companies hire consultants or commit senior personnel far too late, per Figure 2. Typically in the United States when a project starts, the effort is in the spotlight. Management wants funding and credit for launching it, so senior management attention is devoted to ensuring the plan is written and approved. With this completed, there is a lull. Management turns to more pressing duties, and the company's senior professional resource is either used for the next proposal or to fight fires on troubled projects. Finally, with the introduction date fast approaching, the project again becomes worthy of attention. By then it's usually urgent, troubled, or both. It may have wandered from the original goal, or perhaps the situation has changed and the goals need to be modified. Another spurt of expensive resource is applied, but by now there is not much leverage to make changes without suffering major delay or starting over. Sometimes there is even a last spurt of activity long after introduction. If the product is doing poorly, resource may be applied to study or reposition it. By then leverage is very low, so distressful

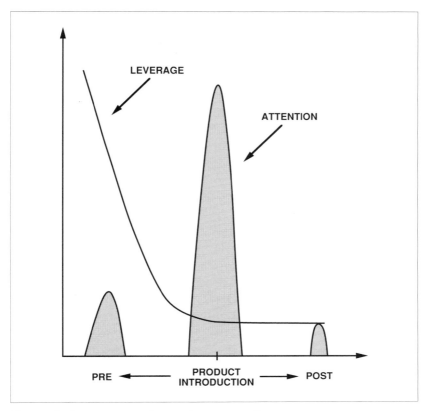

Figure 2. *Leverage and attention versus time.*

expressions like "making silk purses out of sows' ears" may become germane. This is not the best way to run a project, but it is how most projects are run. Wouldn't it be better to apply the resource when there is leverage to cause change, and to leave it in place to make continual improvements?

Regis McKenna "Himself"[2] used to lecture using an analogy. He suggested that creating a technology-enabled product was like launching a rocket to the moon. If you aim at where the moon (the market) is, you will miss, because it will have moved by the time your rocket arrives. In a new market, if you wait to be sure, you will arrive too late to reap the rewards. It is better to launch early based on the best information available. Then you put good sensors and guidance systems in place to allow periodic course corrections. As the rocket nears its target, the sensors can collect a clear view (better understanding and lower uncertainty), and the guidance system can make efficient corrections without expending excessive fuel (money). In your skunk works these sensors and systems are the minds and skills of the "skunks," so you had best get the best available and let them function until your expensive rocket (the product development) reaches your target.

It all starts with selecting the right people and putting them in a structure that allows them to focus on results. The structure is the skunk works itself, and *you should separate it*

---

2 "Himself" is the title on his card. Some thought this was arrogant and perhaps so, but in another way it was equivalent to the "Guru" title I will use later in this book.

from other corporate issues and intrigues. In small companies the skunks are most often the founders and their backers. In large companies skunks can be found in task forces, divisions, or business units, but these are often hostile environments. Increasingly, big companies are trying new business elements and product launch teams, but these are substitutes for a skunk works. The large "industrial age" technology companies are frantically flattening their organizations to allow rapidly moving teams, but they are having trouble doing this well.

The commitment

Starting a new business or moving into a totally new product-market area is hard work. There will be a time during the effort that all seems hopeless, and most likely there will be more than one such time. Many years may pass before even the best new ventures are profitable and can stand on their own, and you should be realistic and accept that. Leadership and vision is required by those with authority to commit corporate resources. If your company's upper management is not firm in purpose and resolve, success is not likely. If you can't tolerate a loss position for five to eight years, then you should not contemplate playing in the "suicide square." (A portfolio approach is a good way to manage risk and secure the cash flow

> "These ventures (in large companies) are relegated too far down the line. They are run by someone who's scared to death of making a mistake because he knows it could threaten his corporate career. On the other hand, if he does well he won't really be rewarded, because what he does is not part of the corporate charter."
>
> — *Reid Dennis*
> *Venture Capitalist (VC)*

you will need to fund your ventures. This will be discussed in a later section.)

It's easy to destroy a skunk works, since skunks are invariably resented and viewed with disdain by the more respectable corporate animals. The most sadistic way I know of to annihilate a skunk works is by the corporate equivalent of "death-by-a-thousand-small-cuts." If you want to ensure that your new ventures fail and endure lingering death, just make it generally known that they are being funded by diverting money from the product divisions. ("You know — the ones that

---

Why are those skunks so impractical?

In the early 80s a large electronics company set up a computer research laboratory, staffed it with excellent people and equipped it with expensive tools. The technologists complained they needed better tools to experiment with. Everyone wanted a computer, but to buy each a mini-computer was too expensive, so the researchers built their own. They did it fast and cheap, and they used standard parts. The machine was named after a flower, it ran UNIX, had a good graphic display, and used a standard Motorola microprocessor. First a few were built, then more, and finally having one became a status symbol. There were soon over twenty prototypes running, networked together, and in daily use.

The director of the laboratory showed the creation to his Vice President, and finally to the board of directors. He wrote a memo suggesting that some minor improvements be made for producability, and that one of the company's divisions try to sell a few to see if there was any outside interest. This could be done inexpensively and in only a few months. The answer came back from above, "Good idea!"

Upper management got excited about the new opportunity. A division was asked to develop this promising new device into a product, and new engineers, marketing people, and managers were hired. They all had ideas to add value. Performance would be better if custom chips were developed, the displays could be greatly improved, and, in fact, an entire family of products was possible. The creators and the implementers soon conflicted, so

---

> "Managers would come to me saying their 'charter needed expanding.' Anyone who says that to me is going to be the sorriest son-of-a-bitch you ever saw. The 500 pound load they're carrying isn't enough? Fine! I'll add another 500 pounds."
>
> — *T.J. Rodgers*
> *Leader, President, and CEO*
> *Cypress Semiconductor*

make money for the company.") This technique is particularly effective in a business down turn, and most large U.S. companies shed their ventures in such times. If you merely wish to castrate the endeavor so it doesn't bother you and steal resources from your product divisions, a good technique is to entangle

the large product organizations took control of the project. The division was soon expanded to a product group, and, since this was very important, an up-and-coming Senior Vice President took responsibility.

The company decreed that all the other divisions would standardize on the new products, so a corporate strategy was developed. Committees were formed to study standards and architecture, and the deeper they looked, the more opportunities they found for technical creativity. Soon over a hundred engineers were working on this project, then two hundred, then three. Years passed and expenses mounted, but convincing presentations were given. The first competing products appeared. They were from tiny new companies with names like "Apollo" and "Sun." The company added more features and performance to their planned products. The number of engineers assigned increased again, but then the next generation of competitive products was released. These were more formidable, and there was a better wave coming behind them. Top management grew concerned, the V.P. was replaced, along with key managers. Another V.P. took over and announced the products, but he soon pulled them off the market when customers complained that they were not finished. The new V.P. was replaced, and the product line was canceled.

Had this been done as a small, fast, skunk works activity, the company would have invested much less and might have entered the workstation market years before anyone else. I think their chances of success would have been better.

it in discussion of "charters" and "strategic relevance." New ventures always consume funds, and new things are invariably too fuzzy and ill-defined to be considered strategically relevant. It took over a decade before lasers or the transistor were strategically relevant, and IBM ignored PCs for almost as long. The future only becomes strategically relevant to a company when its staff moves down the street and forms a company that enjoys success, grows large, and deprives the parent of its revenue base by doing some product they refused to fund. If you force a new venture to focus its energy internally discussing organizational charters, it won't have

---

Who should you bless, and who should you blame?

A friend managed a new venture for a large corporation. With a tiny team, he developed and launched a clever workstation for object-oriented software development. Competition in the field was intense, but this venture was growing nicely. It had parity market share with companies that had committed tenfold the resources. Like most new start-ups it was unprofitable, but the losses were insignificant and steadily improving. Unfortunately, it was set up to report to a Group V.P. alongside several well established divisions.

The new venture was small and puny compared to the large divisions. Worse, the divisions all wanted to put constraints on the venture to prevent its cannibalizing their profitable businesses. For example, the venture was forbidden to host standard operating systems or languages (such as UNIX, C, or even FORTRAN) and this denied much "pull through" revenue for them. The innovators fought off repeated attempts at absorption. It would be more cost effective for the company to merge their manufacturing, or their customer support, or their product marketing. The mature businesses could do these things better, cheaper, and with more finesse. I felt sorry for them, since their management was forced to expend most of its energy on internal issues.

The venture had a small team of specialists to sell its products. One day the Group V.P. agreed to consolidate the sales forces

---

enough left to get its job done. If upper management is not firm in its support and protection, it is unlikely that others lower in the company can have enough impact to get results.

Investment in the future and moving to address new things effectively requires considerable courage and commitment by the CEO of a U.S. company. Investment in new things will have a negative impact on next quarter's balance sheet, and it will never pay off unless you patiently invest for several years. You can bet that (if you allow it) each of your existing organizations will suggest dozens of "more

---

"for better effectiveness." My friend called, distraught, knowing this was disaster. We rallied to his aid. The manager of the laboratory that developed the technology, and the company's V.P. of technology appealed the decision. For this we had to go to the Executive V.P. of the company and then assemble the members of the company's board that were technology knowledgeable. It took months to arrange this, but finally the meeting was held. After several hours of debate, the council decided to allow the new venture to have its sales force back.

My friend and the laboratory manager were relieved, "We won!" I was more cynical, and asked, "How many of your sales people are still with the company?" "About half," was the reply. "Of these, how many would want their old jobs back, knowing how vulnerable they were?" "About half," was the answer. "Do you have the same sales forecast, or can you change it?" "We have the same forecast." I told them that we had not won, that we had lost. They didn't believe me, but within six months there was a large variance between forecast and performance. The venture was canceled, and those managing it were blamed for the failure.

Comment: It's hard for large organizations to move quickly toward uncertain goals, but they politically overwhelm small teams. The "big dogs" have more clout, and management is usually obligated to back them in "charter disputes."

effective ways" to spend your funds. Another
check list is helpful.

☐ Does the corporation require prior proof
of success?

☐ Is the team forced to conform to the company's standard management and administrative system?

☐ Do you insist on only home runs? Is the
team directed to swing for the fences and
promise breakthrough results?

☐ Is funding and direction unstable or inadequate? Must the team compete with

Why are some skunks so hard to kill?

A skunk works may have saved Sun Microsystems. In the late
1980s, Sun was firmly in the lead in workstations. When a proposal was made to improve their products by making them
smaller, less expensive, and more powerful it didn't seem a
prudent business move. In January of 1987 a young engineer,
Mr. Andreas Bechtolsheim, had his proposal for a desktop computer called the Sparcstation firmly rejected. Careful and rea-
soned business analysis showed that the product should not be
done. Presumably it would have disrupted and cannibalized
the sales of Sun's other highly successful products.

Mr. Bechtolsheim was a co-founder of Sun, and consequently
he is independently wealthy. He started a project, and spent
$200,000 of his money on a prototype. He formed a new com-
pany called UniSun to develop this, and hired one of Sun's
board members (a Mr. Khosla, who was Sun's founding President) to be its President. After consideration, Sun decided to
buy the fledgling company and set it up as a division. Having
made his point, Mr. Bechtolsheim generously allowed them to
do this for cost.

That did not end the matter. Sun really did not want to do this
product. They wrapped the project in restrictions, and the division was only allowed to hire those Sun engineers who said
they would otherwise resign. Mr. Bechtolsheim recruited anyway. The project drew strong criticism from Sun's core divisions that made bulky and expensive workstations, but the

large or mature organizations for resources?

☐ Is the venture deemed "strategic" to the corporation?

If any of these boxes are checked, expect failure. I expect some by now are ready to throw this aside in irritation. What of risk management? What of accountability? What of responsibility to stockholders? Why should an elitist organization be permitted to waste scarce resource? These topics will be addressed later, but much of the answer is **TO KEEP IT SMALL AND QUICK!** If your new

team stayed fixed upon their vision of a small desktop unit. In April 1989 Sun introduced the blazing fast Sparcstation 1. It fits on a desktop, and it dropped the price for this level of performance from about $25,000 to $9,000. It was an instant success and quickly became Sun's best selling product. They didn't stop, but aggressively followed their success with improved products of even better value. By May 1990, two more Sparcstation models had been launched, and the cheapest was selling for $5,000. Sun doesn't break out product sales, but admits that the Sparcstation product line is their biggest seller. They should! Vicki Brown of International Data Corporation estimates that it now comprises almost 75% of Sun's total unit sales. Mr. Bechtolsheim prevailed, and is reportedly somewhat of a folk hero at Sun. At 34 years old he is Sun's chief computer designer, and the **Wall Street Journal** reports that it will be a long time before Sun again ignores his advice.

Comment: Mr. Bechtolsheim has now led creative revolution twice. That is commendable, since it is easy to move in short order from being the dynamic young skunk works leader to becoming a conservative member of the corporate establishment. I think he is astute enough to recognize this, since he has reportedly structured his job content in a manner where he lacks budgets and organizations to protect. In a few years, the Sparcstation products will be replaced by something new and innovative from another skunk works. The replacement may again be one of Mr. Bechtolsheim's projects.

idea is to fail, you want it to fail quickly and inexpensively. Even when you are highly successful, trials into new areas may fail often.

You will note that a very consistent theme is the hostility shown by the status quo to the new. This is human nature, not coincidence. If you are successful at what you do, change is threatening and uncomfortable. Creation of the future disrupts the present, and it is rare for the leader in one generation of technology to lead the next. Only the very best leadership can make this happen.

None of the leaders in vacuum tube technology (RCA, GE, Tungsol, et. al.) moved to leadership in transistors. Few of the leaders in transistors were successful at integrated circuits. The best mechanical calculators were developed in reaction to electronic units, but it didn't help. With one exception, none of the leaders in 8 bit PCs were major factors in the markets for 16 or 32 bit machines. Because of heroic efforts from a separate skunk works, IBM finally got into the PC race and prevented Apple from taking the market. In most niches, they missed. IBM never has done much of anything in low-end PCs, laptops, minicomputers, or supercomputers. This kind of behavior has been true throughout history. The railroads and steamship lines missed the next waves in transportation, and few of the leaders in propeller-driven aircraft were at the forefront of the jet age. Military history has collected such examples for centuries. The same tactics that succeeded against muskets were tried against machine guns with tragic results. As late as World War II a few cavalry commanders still preferred horses to tanks, and a few air forces were still equipped with biplanes. General Billy Mitchell was court-martialed for advocating the Army's transition to air power.

Our reflexes and instincts were developed for a time when the world was more stable. When under stress, the tendency is to do what worked last time. This survival mechanism is dysfunctional for success in high tech.

The question you should ask yourself is this. When your products are inevitably replaced by the next wave, do you want to cash out or stay in the game? If you want to cash out, admit it and retire; but if not, be prepared to invest and adapt. I have been told that many leaders of the U.S. automobile industry in the 1960s were thinking retirement. If they had developed smaller cars, they would have taken a risk, and their company's profits and their bonus pay would have been less. If you want to stay in the game in technology-enabled business areas, the key is not so much sticking to your present products as it is satisfying the changing needs of your markets. When your customer's preferences shift to new alternatives, they can either buy from you or from someone else. You will always lower your short-term profits if you improve and replace your products, but you may suffer severe hardship in today's markets if you don't.

... with Low Risk

Funding and control

You don't need large sums to get to a prototype, but it is crucial that you have adequate, stable, and patient funding so the team can concentrate on the task. At the time the plan is accepted, *all* the funding to get to market should be committed and set aside. Release of the funding should be contingent on meeting agreed to milestones. Sample milestones might be technology demonstration, working prototype, alpha or beta tests, and first production. The temptation is to press on with inadequate or piecemeal funding, but I think it's better to stop or do something else until enough money is committed. The real expense comes after the prototype, and if you can't get full funding commitments with a prototype and patents, something is wrong and you should stop or take your venture elsewhere.

Conversely, the allocated funding should be perceived to be the maximum, and there should be rewards for using less. This point is important, so I will repeat it. When the project funding is consumed, it's gone, and everyone on the team should know and believe this. One of my pet peeves is game playing with schedules and funding, but this is common. It doesn't add value for management to arbitrarily cut laboriously planned budgets or schedules. It should be a firing offense for teams to promise what they know can't be done. Schedules and budgets must be deadly serious commitments that are adhered to, and it's irresponsible for management to cut them or for the project team to ignore them.

The timing of money is absolutely critical in high tech. You should have the funds available in a timely manner. The team can never recover the time wasted waiting for funding decisions. You should estimate the approximate amount of money before the plan is prepared and have it waiting. As Eisenhower once said, "The value of a plan is the plann*ing*." The planning process should be used to force discussion, provide focus, and resolve details. The plan itself is a promise, and it should be signed by the team. They are saying, "For this amount of investment, we promise you this revenue stream." If you invest the money, you should make the team accountable for the results. If, however, you take months to study and approve the plan, the delay will make their jobs much harder and it may cause failure. In the 60s and 70s it reduced corporate risk to study plans carefully until all aspects could be "proven." That method does not work well in the 90s. In today's fast moving niche markets, the entire competitive situation could change by then. With today's rapid communications, just working with leading-edge customers on a technology-enabled product changes the market. The best high-tech companies can decide and commit resources in days.

With funding comes controls, and I think they are necessary. They should be administered in real time by the team leader, and the project status should be reviewed periodically, perhaps at milestones, by those providing the money. It's false economy to skimp on talent or tools, so you should not cut corners there. Since all your cash flow in a skunk works is negative, you must be frugal. For one thing, if you force the team to "work fast and cheap" they will automatically focus on the key areas and reuse existing components and subsystems. This not only saves you time and money since they won't be "reinventing the wheel," but it may reduce the risk of design defects.

Fortunately today's technology provides leaders with the tools to monitor progress in flat organizations without consuming excessive time. You need to manage results, so each person should have time-committed objectives. Management by objectives (MBO) is the technique, and the objectives should be formally agreed to by the team and its management, not set by edict. The objectives must be real, measurable, realizable, and significant, and care must be taken to limit this to a short list of items that truly show that progress is being made toward the agreed upon goals. Schedules slip one day at a time and surprises cause devastation, so MBO progress should be monitored weekly. It's worth spending an hour a week to monitor progress and adjust objectives. I've had researchers ask, "How do you reward the progress of someone half way through a multi-year program to find a cure for cancer?" My answer, "By measuring how realistic their goal setting is, and how they perform against these goals."

You need controls for product planning, head count, capital spending, and expenses. The trick is to make these systems simple, consistent, timely, and just. There is no point in trying to describe the best system, since it doesn't exist. Tools that help one organization or culture may not work for another. The main thing is to avoid bureaucracy and administrative activities that divert energy, but to preserve real-time control of progress.

Paradoxically, I think communications and slack are also needed. I like to set aside a fraction of time for unplanned "pro bono" activities, and even tiny organizations can do this. A good guideline is seven to ten percent. Time commitments within these limits should not have to be justified, but results should be periodically reviewed. Communication is key in the information age. The U.S. Navy has a legal command, "Pass the word." All team members should be held accountable for communicating. In times of high stress or low overlapping objectives, I've used a human resources group facilitator to force agreement of headstrong team members to formal CAIRO documents. These are lists of who must be **C**onsulted, who must **A**pprove, who must be **I**nformed, who must **R**eview, and who can be **O**mitted concerning actions in specific areas. Techniques such as this can be helpful when several disparate technologies must be brought together to provide a product.

> **Note:** I chose to use a CAIRO document in this case because my style is participative, many people did not report to me (not the best way to set up a project, but it happens), and I've found that peer pressure often works well on technologists. I think there is value in forcing the team to become self organizing. Another leader might choose to be more directive in such a situation, but the point is that the

leader must set limits to manage direction and conflict without stifling creativity.

The right talent

Y ou should know precisely what you are looking for if you hope to find it. I distinguish carefully between marketers and salespeople, creators and implementers, and managers and leaders. This is more than semantics, since these can be very dissimilar people. My point is not that some categories are better than others — you need them all. It's that you had best know what resources you have, who is superior at each skill, and when you should apply each. The main assets of a knowledge-based business are people, so managers should be *personally* responsible for recruiting, hiring, and retaining good talent.

The leader

The hardest role to fill is that of the sponsor and leader. It's the one where outside talent cannot help you, so you must find a leader in your organization and allow them to function with a high degree of autonomy. This person has to form the team and provide a nurturing, supportive environment so that they can function. They must, unfortunately, do this through all business cycles, with no useful quantitative short-term measures of success, and while consuming capital and "failing" more frequently than any other part of the company. The leader must accept — and *survive* — frequent setbacks. The economic return from a skunk works or a new venture comes years or decades later and mostly accrues to others. (There is nothing malicious about this, it's just a fact of life. Profits and revenue will peak years later, and by then the company will be selling a second or third generation descendent of the original product. By then the skunk work's contributions will be dim and distant memories, and the innovators will be working on something new.)

Comment: The following net present value calculation is informative. Choice a) is to make $200K per year, and increase it by inflation for the next seven years. Choice b) is to make $100K per year, hold it constant, but to have a 10 percent chance of making several million dollars in seven years. The numbers would be higher for California and lower for Oregon, but a) is a general manager in a mature company and b) is the CEO of a new venture. The present laws tax both revenue streams at the same rate, so which would you prefer?

A leader must be dauntless, dedicated, and motivated to endure such a role and such a reward structure. The best leaders are often found in front of their troops, so the risk of harm from "friendly fire" is high. It's common for visionary leaders to leave (or be forced from) the successful companies they started and be replaced by "professional managers" who generate more short-term profit. It's safer today in the U.S. to build corporate careers on paper entre-

> "Americans today make money by 'handling money' and shuffling it around, instead of creating goods with actual value. America needs to get back to a real production economy. A quick profit from a stock deal should be taxed at a higher rate than those on long-term capital gains."
>
> — *Akio Morita*
> *Leader*
> *Sony Corporation*

> "I decided at (a growing Fortune 100 electronic company) that I wanted to start my own company. I was tired of spending 40 percent of my time in politics and turf battles."
>
> — *Roger Ross*
> *President*
> *Ross Technology*

preneurship and cost cutting than on leadership and the future. Foreigners are appalled that American financial markets roll their investments in minutes (some traders have ten-minute investment spans) versus their horizons of ten years or more. U.S. corporations employ many more managers than leaders, and it's easier to become a Vice President by not making mistakes than by taking risks. It's easy to generate short-term profit by selling assets, deferring maintenance, reducing R&D, or killing new ventures. It's very easy in large U.S. corporations for managers to become wealth centered rather than opportunity centered. Wealth-centered managers tend to look to the past and protect what they have. This soon leads to a "downsizing mindset" where managers tend to focus on minimizing their personal risk and maximizing short-term profit. This is an unhealthy game where the decisions amount to driving toward removing as many assets as possible from business endeavors, and the only limit is to avoid killing them during *your* tenure — unless someone else can be blamed for the death.

Anyhow, you need a true leader if you hope to be successful at new things. What type of leader you need depends upon your culture. I personally favor a participative leadership style for high uncertainty efforts, but others have good success with authoritative methods. The Japanese exemplify the former, the

Germans the latter, and there are successful examples of both styles here in the United States.

We've talked about motivation for the leader, and how he or she must protect the venture and secure its funding.

> "A leader is best when people barely know he exists. When his work is done, his aim is fulfilled, they will all say, 'We did this ourselves.' "
>
> — *Lao Tse*
> *Writer (600 B.C.)*

Other than that, what value does the leader add to the team? Managers handle or manipulate, but leaders *lead*. The leader sets direction and limits. The leader chooses the goal, and by all forms of communication, especially by example, shows the team how to move effectively in that direction. Leaders shepherd, direct, arbitrate, coach, counsel, and serve as the team's resident psychologist. They select the team and occasionally they kick ass and fire people. The trick is not to get people to do things they don't want to, it's to get them to *want* to do the things that lead to project success. Leaders set the project limits and ensure that neither the marketing Guru nor the technical leader allows the "creeping feature creature" to devour the project. There's another thing the leader must do, and that's to orchestrate and serve as cheerleader to aid team building. Celebrations of accomplishments are important. When your team finally works as a team (this takes time and mutual respect) you will accomplish wonderful things.

Who should you choose to be the team leader? You should not select a staff person or a sycophant, but neither should you select a "wild man" with all thrust and no rudder. You need someone whose judgement and ethics you know and trust, and someone with the courage and decisiveness to take independent action. You need someone trained and seasoned, and someone who has a good track record of success in technology-enabled products. Above all, you need someone willing to lead and able to inspire others to follow. It is absolutely critical that both you and the key team members can explicitly trust the person. If you were part of a two-person team climbing a sheer rock cliff, would you want this person to be holding your safety line? Would the others on the team select him or her to hold their safety lines? If you had to send someone on a long and arduous journey with a large sum of cash to ransom a loved one from terrorists, would you select this person? If you had only one chance to give a presentation to save your company by explaining the technical, financial, and market appeal of your new project to a panel of experts (and could not do it yourself), would you choose this person? Would this person use money and power to create and grow, or to abuse and exploit? I personally would never select anyone who lacked experience at fast-track technology-based projects. Good candidates might be the technical and marketing leaders of previously successful fast-track projects, or someone who has shown the courage to leave a large corporation to pursue a venture independently.

> **"I want my people to take risks and to be creative. I expect them to fail often, but I want them to fail fast and cheap."**
>
> *— John Stoltz*
> *President and Leader*
> *Advanced Systems*
> *Concepts, Inc.*

What if someone has failed at a technology-based project or venture, should they be a candidate, or are they a "turkey for life?" Most of today's venture capitalists would brand anyone who failed at a venture a turkey for life, but I think they are misguided to take this attitude. Many

> **"Imagination is more important than knowledge."**
> — *Albert Einstein*

exceptional people have failed, but learned from the experience and were the better for it. I think good leaders should encourage promising people to "push the limits" and allow occasional, non-catastrophic failure. Some of my best education has come from studying failure, sometimes my own. If someone fails, you should ask "Why did they fail, and what did they learn?" Did they fail based on a fatal flaw or a personal innate weakness, or did they just make a mistake? Do they have a winner or a loser mindset? (Some people fail because they don't know when to quit in the face of hopeless odds. You might want such a person.) Can you have your team leader candidates "practice" on productive lower-risk projects before you move them to one in the suicide square?

You say you cannot think of anyone that meets these leadership qualifications? Sure you can. I can think of a dozen names off the top of my head, and some are mentioned in this book. The problem is that you likely can't induce these people to work for you. If you want to compete in the technology markets of the 90s, you had best start finding good leaders or growing them yourselves. Leaders, like eagles, are hard to find, and you never find them in flocks. There is a reference on leadership in the

> **"The future is the past in a different context. The problem is trying to understand this and take advantage of it."**
> — *Phil Crosby*
> *Chief Scientist and Inventor*
> *Tektronix, Inc.*

bibliography. If you have no strong candidates, you may have to run the venture yourself until you can groom one. Staffing is the hardest thing a manager can do, so you might wish to use first-rate consultants to help you decide.

The technologist

Winning today depends not on resource level, but on enabled minds operating from timely and correct information. Much has been written on the technical part of the equation. How do you find creative inventors that can produce practical results? Frankly, there is no known formula for selecting genius. Some first-rate technical people I've known lacked formal degrees. I recall a brilliant antenna designer I worked with long ago. His peers worked with computer models of Maxwell's equations, but he was a music major, drunk during his off work hours, who created shapes that felt right. His designs worked consistently, and better than theirs.[3] Genius is crazy, almost by definition. To win in high tech you have to imagine what can be, and to implement rapidly based on intuition. If you wait to be sure, you will fail.

Good companies invest heavily in developing and grooming their technical talent. At the bottom, there is usually a recruiting process to hire the best students from good universities. Next there are grants to encourage graduate study, and to induce these students to work for the company after finishing their degrees. On-the-job mentoring is common, as

---

3 Gilder's **Microcosm** talks about this phenomena, and his chapter on "Analog People" is a classic.

are both formal and informal apprenticeship programs. In these, the engineers are assigned to work under more experienced technologists, so that they may practice and learn. Companies go to great pains to provide information exchange and cross-training, as well as training in the engineer's core area. Design audits are a common tool, as are internal workshops, technology fairs, and seminars. The design audit not only is a good way to assure product quality, but if done correctly, it can provide useful training.

Whatever the name of your company is, most of the excellent talent in the world works elsewhere. The very best companies figure out ways for their engineers to develop close contact with this outside talent and gain knowledge. Committees and trade associations are some techniques, but there are many others. Awards for achievements are common, and good employers encourage their engineers to publish and gain professional recognition.

Hewlett Packard develops their engineers as well as any U.S. company that I have seen. They do everything mentioned above, and a number of other clever things. They often link the publication of significant technical papers to their product releases. If, say, they are going to announce a new digital oscilloscope, the first things that appear are technical papers on related but esoteric subjects. Perhaps a paper will appear on a new process for integrated circuits, or on research into a new digitization technology. HP's researchers then go around the world and present their results at conferences or seminars. This is a very smart thing to do, because their engineers not only get the best possible peer review and contacts, but they also generate interest and respect for the company's technology. After some delay, the product is announced, and promoted. This is an intensive event, with professional PR, press releases, beta-site testimonials, press tours, and advertising. Some engineers participate and learn. This is followed by technical articles in the trade press, telling how the implementers designed the innovative product. The implementers now get their wave of external peer reviews, exposure, prestige, and industry interest. Still later, HP's application engineers start publishing on how to use the new product effectively. The cycle repeats, and more interest, more exposure, and more training result. Over the course of a year or so, many things have transpired, all good. HP's engineers have gotten absolutely world-class technical feedback, and they now are smarter and more seasoned. They are also starting to become recognized — those that are not already — in their fields, and they have an expanded network of contacts to help them develop and perfect new ideas. (Their marketing and sales folks are as happy as clams, because their product introduction has been stretched from a single event to a year-long relationship, and the next product is about to be announced.)

Every high-tech company does its best to retain and grow technical talent, and discussing the care-and-feeding of engineers is a topic for another book. If you do not have a good pool of technical talent, you should not be attempting to compete in technology-enabled business areas. This is as basic as having desks for your workers, so I will assume that you understand, have done this, and press on.

There is another important detail. To become good engineers, your technologists have to not just train, they actually have to do successful products. I had a good friend who said that advanced development was, "Where the rubber meets the sky." Whatever, if your engineers do not practice their craft, they can never become first-rate talent. This, more than anything, may be the phenomena that built Silicon Valley. Technology business there sizzles, and the local engineers develop an incredible pool of shared and current knowledge. The engineer's loyalty to an employer lasts only as long as they are effectively using his or her talent. To prosper — perhaps to survive — engineers must "stay current." If a company has poor management and starts to lag, a leading indicator is that their better technologists move down the street to work for someone else. (The financial community might do better if they tracked this drift of talent to forecast high-tech corporate performance. This is, of course, heresy, and even worse it implies a long term interest by the financial community.) Talent drift in Silicon Valley happens rapidly, but the engineers in the Northwest are more tolerant of bad management because it's a good place to live and there are fewer places to work.

My personal bias is that engineers should be educated and trained lavishly, but they should be paid based on their contributions to successful products. It's really quite simple. If they are unable to create successful products at your company, your best talent will eventually leave. If they can create successful products, you should pay them well. Too many large companies drift into mediocrity. They retain large technical staffs, and they pay the average wages needed to keep them. They have their human resource departments collect data and hold salaries down by persuading the engineers that they are being paid competitively. There is usually nothing wrong with the technologists in such companies, but neither is there anything exceptional. If I were King, I would employ a smaller staff of exceptional talent, but I would compensate them well. I would spend about the same amount of money, but I would get better results.[4]

Companies with a rapid cycle of product introductions enjoy many advantages. One subtle advantage that is not often discussed is that their engineers are better — because they have more relevant current experience than their peers. Consultants, if they make a living and avoid burnout, can enjoy (?) the most current and intense experience of all. A friend is Board Chairman of one of Oregon's larger and more successful companies. He told me that he has suffered grievously in the past due to this phenomenon. The design groups for his most successful hardware company are now located in Silicon Valley. He found that the engineers in Oregon were good conceptually, but could not match the implementation experience of their California peers. Another friend manages a smaller Oregon hardware company in a fast-track market

---

[4] I must credit Robert Townsend, the rebel-with-a-cause who turned Avis around and wrote 1970s number-one best selling book **Up the Organization**, for teaching me this simple but rarely practiced fundamental of business success.

area. His company has excellent products and has been running neck-and-neck with a few competitors located in the larger technical centers. Unfortunately, the pace is increasing, and I doubt they can fund the rate of product development needed to stay at the front of the pack. He has a great team of implementers, but his competition is good too and has deeper pockets. You need both money and talent to win in high tech.

> **Comment:** Our relative lack of infra-structure is Oregon's curse, though consultants and small software companies can maintain the pace needed to stay at the forefront of their industries because we don't need much working capital. Although almost all our business is elsewhere, we can easily commute electronically and otherwise.
>
> With a few exceptions, our local brothers and sisters in the hardware sector have much more severe problems. They especially have trouble getting the funding needed to run with the fast movers. In Oregon we can raise modest amounts of money to pay for capital assets, such as buildings and lumber mills. Unfortunately, most high-tech companies need working capital, not capital assets. It's difficult for an Oregon company to raise the levels of funding that a company in California, Texas, or Massachusetts can. Fast growing hardware companies burn cash at unbelievable rates, so if you plan to start a Compaq or a Sun, you

might do better to site it close to your capital. (Intel, Sequent, Mentor, HP, and the Japanese plants largely avoid our infrastructure constraints. Their funding mostly comes from out of state, and Sequent and Mentor both got their venture money elsewhere and during the flood tide of the early 80s.)

Capital access is a major problem for the entire industry, of course. I think the early leader in workstations, Apollo, was sold to Hewlett Packard mostly because it was the only way they had to fund the rapid growth needed to stay competitive.

How do you select a technical leader for your venture? Simply list the top five people in the world at creating products in your technology area of interest. (You can't list five? This would be a good time to learn who they are.) Ask each person on the list to make a listing of five candidates, and then correct yours. You now have the choice of either hiring one of these, or growing someone that would qualify. Probably you will find that you have no one that qualifies, and that means you are early in the research phase of the project. Do not get discouraged. Get the best talent available and have them start gaining knowledge and experience. Our Japanese and European friends regularly loan their most promising talent to work in places like Bell Labs just to develop such capabilities. Again, you can train your young talent by having them start in lower risk areas than the "suicide square." It's sometimes effective to use a good consultant or share a chief engineer for your technical leader, *if* you back them up with good apprentices when the project expands.

The marketer

So much for the technical part of the team. Find a creative genius or two, and retain them. You have to look long and hard for them, since they are disparate and unlikely to make it through the screening process of your personnel departments. The same is true in the area of marketing. This causes high-tech companies endless grief. Marketing is the weakest link in most technology-based U.S. companies. Even the definitions become confusing, and you must learn to distinguish between sales and marketing if you hope to succeed in high-technology business. Sales or promotion and marketing are very different things. Sales is inducing someone to buy what you have, and you normally do this after product introduction. **Marketing is identifying a need and filling it profitably, and you must start this *long before the product is defined* and continue it long after the launch.** The skills are not necessarily interchangeable, so salespeople usually make poor marketers and vice versa.

Incidentally, even the label "high technology marketing" is misleading. Market*ing* companies focus on technique, on promotion, on sales, and on advertising. Most expectations about marketing come from the consumer sector, where shifting a single share point for soft drinks nets enormous returns. High tech is different, and you will waste

> "Promoting a bad product well will cause it to fail sooner."
>
> — *Anon*

> "Great devices are invented in the laboratory. Great products are invented in the marketing department."
>
> — *William Davidow*
> *Marketer and Venture Capitalist*

much money and time if you don't explicitly acknowledge this. For many fundamental reasons, you can't promote your way to success in high tech. IBM spent over $100M promoting the PC junior, the Edsel of the industry. It's better to be driven by your market than by your market*ing*. You need to think in terms of doing "positioning and infrastructure development" before you do promotion, but these concepts are more often abused than used. Everyone uses the "buzz words." It's common to give lip service to these concepts, but rare to practice them well.

Marketing's main job is ensuring your company, its products, and perhaps the company's technology are positioned for maximum competitive advantage.[5] It's rare for marketing in a high-tech company to be a full partner in the product development process, but over 60% of high-tech product failures can be attributed to poor market targeting.[6] It's not good to make your engineers responsible for marketing. It's dangerous to direct them to "talk to customers" to justify working on their projects. Customers are polite, and generally say kind things about new product ideas. The usual conclusion is that it's good to improve products and to provide more features and

---

[5] There are many other ways to differentiate your offerings and provide competitive advantage besides product features and attributes. I will not dwell on these, since they are covered by any good marketing text.

[6] Source: TTG research.

more performance at less cost. Such a dialogue is not very helpful. You can never be sure if the engineer's "market research" centers on interesting

> **"If Edison had been an MBA, he would have invented a large candle."**
>
> — *Anon*

technical problems or business opportunities. Most likely it was the former, because that's the primary focus of most good engineers.

> **Comment:** U.S. technology companies frequently say "marketing is everyone's responsibility" to excuse their weak marketing functions. Strangely, they rarely say that "engineering is everyone's responsibility," or "manufacturing is everyone's responsibility." The problem is that everyone's responsibility turns out to be no one's responsibility. Someone must be accountable for results.
>
> An even stranger theory is to try to make your customers responsible for marketing. After all, don't they give you the best ideas for new products? Unfortunately, it's your job to sort out what customers tell you and make the right decisions. Customers sometimes know their future needs, but they don't know your business as well as you do. (If they do, something is wrong.)

A good litmus test to tell whether you are doing market research or just holding customer discussions is to ask the following questions. Are the features and specifications of your planned product changing in unanticipated ways because of your efforts? Have

> "The fellow who can see only a week ahead is always the popular fellow, for he is looking with the crowd. But the one that can see years ahead, he has a telescope but he can't make anybody believe he has it."
>
> — *Will Rogers*

you discovered anything that you did not expect to find? If not, you either miraculously conceived the product perfectly yourselves, or, more likely, you are not listening. You should hold marketing accountable for creating new products, but give engineering the credit, responsibility, authority, and accountability for implementing them properly and quickly.

The objective is to meet market needs, but this may not be what your technologists want to work on or what customers request in surveys. Useful knowledge takes deep customer knowledge, an intimate understanding of technology, and a close working relationship with engineers. There was no market for personal computers or LANs or laser printers or spreadsheets until the product existed. The best technology markets are creations of the mind, and it takes as much skill and art to create these as it does to invent the devices.

Don't think that somehow your marketing Guru can take an arbitrary technology and magically invent a market for it. That's *not* how it works, and many clever technical ideas will never become profitable products. There has to be an unfilled need that customers can be economically taught to recognize, acknowledge, and value. The customer must like the product, under-

> "Beware the 'gold toothpick' type of product. Every customer you ask would like to have one, but few would pay enough money for you to profit."
>
> — *Lew Terwilliger*
> *Marketing and Sales Manager*
> *Tektronix, Inc.*

stand it, and be willing to pay money for it. The trick is to be shrewd enough to imagine the latent opportunities available to you, and skillful enough that your detailed implementation meets customer needs profitably and in a manner that gives you competitive advantage.

It's at this point that my reviewers inevitably suggest, "Expand on how to find out the customers' needs." If only it were that simple.

Does the traditional approach still work?

One of The Trudel Group's (TTG's) clients recommended us to a friend who was President of a small California company. The company consisted of a clustering of particularly creative technologists. They were doing well, but for several valid reasons thought the area they were in had a limited future and wanted to diversify into software product areas. They lacked experience with such products, but were investing heavily and betting the future of their company on being successful. They were confident about their technical ability, but in a quandary about what to do to make their product successful.

The President agreed that their greatest risk was identifying their market and ensuring that their entry product had the right features. The first few discussions went well, and he finally said that he was going to suggest to his staff that I visit to investigate developing a business relationship. After discussion with his staff, he changed his mind. He was friendly, and took the trouble to explain their reasons. Clearly they had discussed this issue earnestly before coming to their conclusion, and their logic was impeccable.

If they hired me I would charge for my time and travel, and this was money that could be spent writing code. Their staff lacked the proper expertise to select a marketing person, and would waste their company's money if they picked the wrong one. Selection was difficult, since even someone with references and a good track record might not do what they wanted. These points were troublesome, so what should they do? Since they could not be assured of results, it might be better to save money. Perhaps they could find someone who would charge less. Perhaps they could even find one who would work free or on a

That's like saying, "Expand on how to invent the transistor or the personal computer." The processes are analogous, since in one case you are inventing devices, and in the other products and markets. You can train technologists in technique, but that doesn't necessarily help them to invent new things. You can train marketing people in technique, but that doesn't necessarily help them to create new markets. In the end the best marketing reduces to

percentage basis. Would it not be better to find a retired executive, a student, or a more junior person? If the person failed it would cost them less, and if they succeeded it would be just as good. (He never stated it, but this approach also would allow the designers to continue doing what they knew was best.)

Comment: This is the classic approach, and an example of the style of project management that I call the "skinny arrow." A more effective project organization alternative will be discussed in later chapters.

The results from this President's decision won't be tallied for several years. He may succeed, but the odds are not good. The U.S. has over 14,000 very competitive software companies, so it is not easy to find a beachhead for a new entrant. It's a tough market. The top four companies account for 60% of all software sales, and they are all young ventures that were formed and learned to compete in the tough markets of the 80s. My guess is that another dozen firms share the next 20%, and the rest fight tooth and nail for what's left. The very best software developers are having severe problems with product targeting. Mitch Kapor's team of software experts went through three years of agony trying to define their product features. When ON Technology's (some called it OFF Technology, since the project had been redirected so often) first product finally launched, over $9M in equity capital and several generations of product plans had passed.[7] The product was somewhat minor, and bore little resemblance to the one originally planned.

---

[7] Source: May 1990, **Wall Street Journal** cover story, "Creating New Software Was Agonizing Task for Mitch Kapor Firm."

intuitive understanding. You have to ask the
right questions to the right people, to listen
intently, and to understand what the real
needs are. This is very hard to do, and those
who work successfully in this dimension have
a difficult time explaining how they get
results. When you look back after success has

Don't good engineers know what to do?

It was very common in the 60s and 70s for successful U.S. technology companies to be "engineering driven." In those days, a
company with accepted brands could ensure success through
technical improvement. Each new product would be smaller,
have more features, and better performance. Doing this usually
made the customers happy. If you were not sure what to do,
you asked the engineer at the next bench if he liked your
design. Brand loyalty was high, and engineers ran the company
and made the decisions. Most of today's larger firms have their
roots in that era, and have fond memories of the good old days.
The companies were successful at the time, were good places to
work, and offered job security.

The comment one grizzled senior engineer who reviewed this
text made is archetypical for this type of company. "In our company's early days we were very successful. We had no (he
underlined the word 'no') marketing job titles in the entire company. Then we decided to 'get sophisticated' in the 70s. We
hired business people, and even marketing people. The rest is
history, the company is now in desperate shape, and many fine
people have lost their jobs." This engineer's view is not an isolated one at his company. Their Vice Presidents have been
quoted in the press as reflecting fondly on the past and saying
things like, "you didn't worry about marketing hype. If you
know how to build a better product, you (had) better build it."
The old model is that good engineers do good products if left
alone, and many (perhaps most) U. S. electronics companies are
still run this way.

Most organisms (and organizations) when placed under stress
do what was successful last time. As stress is increased, they
tend to do so with increasing fervor and desperation. Usually
they do this until they succeed, or die. There is no evidence that
this is an effective survival mechanism in high tech. Despite
why they act, engineers invariably justify their actions with

---

been achieved, it seems obvious. "Of course there's a market for PCs, radar detectors, computer aided engineering (CAE) software, walkmans, or whatever. Everybody knows that!"

Unfortunately, most new things are not at all obvious before their time. Seeing what to

---

facts and logic. The engineer quoted here is sincere, his facts are correct, and his logic is impeccable. I think his hypothesis — that trying to develop a marketing capability somehow *caused* his company's fortunes to decline — clearly is incorrect.

An alternative hypothesis, one equally supported by logic and fact, is that the world changed, and evolution simply brought forth new organisms (and organizations) that were better adapted to survival. There are many additional facts available that fit the second hypothesis, but not the first. The competitors of the company where this engineer works are *prospering*. Some are new companies, but the new leader at the high-performance end of the market was a long time adversary. This competitor has traditionally been excellent at technology. They still are, but they have now evolved and adapted to today's world. Today they also have world-class management, marketing, sales, and production. They took the trouble to develop other corporate strengths that complemented their strengths in design and technology. They never beat the engineer's company at its own game, but they inexorably moved the market to types of products where they had advantage. That is called "corporate and product positioning."

Comment: There is some point in time where improving an old product's technical attributes passes the point of diminishing return. There may be small markets for refined versions of yesterday's products, there may be some things these old products do best, but their time in the limelight has passed. Douglas Aircraft long led the world in airliners. Their finest product was the DC-7. Douglas built the best reciprocating engine airliners, this was the best of the breed, and there will never be a better one. Unfortunately for Douglas Aircraft, Boeing developed jet airliners and took over the market. This wonderful product passed into history. I recently spoke with a senior

(cont'd)

do in technology or market creation *before* it is obvious requires a different dimension of vision, one that few possess. Timing is very important.

> **Comment:** I founded a CAE company in 1970. The products worked and I thought the market opportunity

---

Don't good engineers ... (cont'd)

pilot who flies DC-7s for fire fighting. He has flown almost every large military and commercial aircraft, but this one is by far his favorite. His company is upgrading to turbine powered aircraft that can do the job as well, but it won't be the same.

Sometimes old technologies persevere, but hanging on to the past is not a healthy product-market area for a technology company. For decades there were residual markets for vacuum tube based high fidelity products and germanium transistors. I own a wooden winged aircraft, and the model is still in limited production. It is quite fast, its structure can withstand stresses far beyond what today's regulations require, and a few connoisseurs still argue the merits of Sitka spruce over aluminum. Sometimes technical companies survive in such specialty niches, but most have declined and been purchased by others. Just as competitors can "FUD you" for being the first to use a new product, they do the same if you are the last to abandon an old one.[8] In the case of my aircraft, I endure comments about termites, dry rot, and the like. It's a technically superior product in many ways, but its resale value is low.

---

8   I assume all in technology know of IBM's classic FUD (**F**ear **U**ncertainty and **D**oubt) strategy. This technique blocks sales of your competitor's innovative products by playing on fear. "That product is odd. What if it isn't supported? What if it isn't compatible with other vendor's offerings: What if it can't or won't be upgraded: Can you really trust such a small supplier: What if they go out of business?" See **The Regis Touch** for a discussion.

was obvious, but I was young and naive. It took ten years for the market to be worthwhile, and fifteen for it to be obvious. This lag seems typical of some technology-based markets. It can be lonely for the pioneers. For example, even today, businesses based on artificial intelligence or computer-aided software engineering are still slowly gaining momentum. It took decades for the transistor to turn into a substantial business opportunity.

Regis McKenna has written extensively on high-technology marketing, he writes well, and he also has struggled to describe how one can know what customers need. He wrote a delightful paper "Why High-Tech Products Fail" in an interesting publication called **High Technology Marketing Review** in 1987. (The magazine folded after three issues, probably because of its dysfunctional title.) He started with comments from a book called "The Fourth Dimension: Toward a Geometry of Higher Reality" by Rudy Rucker. Imagine a two-dimensional being that lived as a shape on a plane of paper. Could this being see and recognize its neighbors?

"The initial question," says Rucker, "is the problem of how these lines and polygons can see anything at all." If you were to put cardboard shapes on a table cloth and then lower your eye to the level of the table, all you could see is some line seg-

> **"New-technology products need to seek a new dimension — a dimension that allows customers of technology to see, feel, enjoy, and become excited about the application of technology."**
>
> — *Regis McKenna*

ments. How could you discern shapes? How could you tell a line from a square? Obviously, it can't be done without moving into a higher dimension. Viewed from the third dimension, the patterns are obvious.

> "There are many important things to consider when developing new technology, but the first three are people, people, and people."
>
> — *Kevin Considine*
> *Former V.P. of Technology*
> *Tektronix, Inc.*

"Where's the beef?"

We had been doing a small amount of work for a Silicon Valley client for some time. The client had competent professionals, experienced managers, and investment fund backing. The key product was innovative, and was the creation of the company's founder and board chairman. Customers reportedly liked it, and it addressed a broad range of markets. The company's executive turnover had been high, but they had just raised additional investment capital and they were growing. The Trudel Group (TTG) was asked to take a more active role, to conduct interviews, to study the product, and to make a written report, followed by a presentation to the board. The request was explicit, "We want to hit this hard and make some decisions."

"Do you really want us to do this?" I asked, sensing something was amiss. The CEO said yes, as did the most recent investor, and the board chairman. We took the assignment, did the work, and the results were surprising. We looked into the product's attributes and compared them with the needs of the primary market. We found serious problems, so we moved to the next market. Again, we found a mismatch, and looked at the next. Again, we could not understand how the product could be used to do what the customer needed. Since the product was reportedly selling well, we thought we had either made an error or missed something. We checked our information repeatedly.

Then we looked at the order data and the interactions between sales and engineering closely. Most customers praised the product, but there were almost no repeat purchases. The sales literature and testimonials spoke of concepts and specifications,

Marketing is tragically weak in most U.S. high-tech companies, and the term has poor connotations in the industry. It's rarely a full partner, usually involved too late to impact product targeting, and seldom staffed with first-rate technical talent. We are strangely myopic. We fail most often due to shoddy marketing, but our consistent reaction to failure is to spend more on technology or promotion.

but took great pains to avoid mentioning what the product could be used for. The sales people were frustrated, but the engineers were angry. "We hired these marketing people to sell our product. Why don't they do their jobs?" Incredibly, the company did not have a product although they had been selling something for over a year. The device was so cute that a number of companies bought one to evaluate. It was not very useful, so they rarely bought another. We concluded that the product needed major and expensive changes, and submitted a report with recommendations.

The response was a period of silence, broken only by a complimentary message on our recorder from a mid-level manager. Our calls were not returned. We eventually got a curt letter from the CEO terminating our contract. We were never asked to make the presentation to the board, and it was the last work we did for that client. This was unfortunate, since I think they could have become a very successful company had they made other choices. In my mind's eye I envision a sleek speedboat, powerful engines roaring, but with a large and heavy anchor dragging the bottom.

Comment: The lesson here is technical arrogance, not dishonesty. In the "good old days" if the technologists liked the product it often succeeded. Companies with this mindset expect their pseudo-marketing to sell the product their engineers created, and without excuses. When I last heard about this former client, they were still looking for a market to fit their existing product. They still hadn't discovered one, and they were down to considering third world countries.

Because "high-technology marketing" has become almost an oxymoron, I will formally define the talent we seek.[9]

Guru: The high-tech marketing equivalent of a Chief Scientist or Chief Engineer. A seasoned, senior, technically competent, business knowledgeable, world-class expert. His or her area of expertise is "bending" technology-enabled products and businesses to meet market needs profitably. Qualifications will be determined by training, experience, track record, and the recognition of peers. I suggest a minimum qualification for Guru is to have created new markets for at least three technology-enabled products.

Comment: Our clients sometimes ask if we can train their personnel to become Gurus. Training does little harm and it may help, but there is no "magic wand" to create either technical wizards or marketing Gurus. A good process I have found is to cycle superior design engineers through an "apprentice Guru" program. Reading and study is necessary, but the best form of learning is experiential. Such training benefits many engineers, and some become interested and good enough at Guruing to move to a different career track. Conversely, I

---

[9] I am indebted to Lee James of Regis McKenna, Inc. for enlightened discussions and suggesting I move away from marketing terminology and use the title "Guru."

have had little success at training business generalists to become sensitive to technology leverage points.

Anyhow, you need a visionary and creative marketing Guru on your team. Unfortunately, you have *much* worse problems identifying marketing talent than you do technical talent. It's worse because all but the poorest marketing people are convincing. You can spend much money on marketing, not get it, and never realize what you're missing. Ads can be run, press conferences held, market research conducted, application notes written, trade shows attended, sales calls made, and you can *still* be left with little to show for it. *To the untrained observer, each activity may seem reasonable.* The only clue is that results don't come, but there is always an explanation. When companies realize results are lacking, it's usually too late. Most often in these situations the form is right, but the content is wrong. You're asking the wrong questions, sometimes to the wrong people. The ads are sending the wrong message, the sales people are promoting the wrong features, and the press conferences are discussing the wrong subjects. Saddest of all is that your creative engineers have spent their efforts developing the wrong device specifications and attributes.

You must find and retain a Guru, and the selection or development process is similar to that which you used for finding your technical leader.

> "My complaint about the companies I've seen is that they keep trying to substitute process for market knowledge. They get their forms filled out, but they don't understand the questions that should be asked."
>
> — *Tom Dagostino*
> *Product Planning Manager*
> *Tektronix, Inc.*

List the top two in the world in your market area, have each list two, get the best one available, and let him or her gain knowledge.[10] Since Gurus are rare, seriously consider tapping outside talent. If they cannot *consistently* get you good products, you don't have a Guru and you should replace them. Pay no attention to the trappings. Those best at marketing technique are the people seasoned in the dog-eat-dog consumer product sector, but they seldom grow into Gurus. They become enamored of process, and few take the trouble to understand the subtlety of the markets and the technologies they must work with. Pay attention to results, and mate the Gurus with your best technical talent.

> "The first principle of management is that the driving force for the development of new products is not technology, not money, but the imagination of people."
>
> — *David Packard*
> *Co-founder of HP and Leader*
> *of Business and Industry*

The best technique is to have your Gurus and technologists interview leading-edge customers together. The Guru seeks opportunity, and the technologist solves problems. When they decide what to do, make them write a business plan (with milestones and budgets) for approval, and hold them accountable. Then let them run *together* without undue interference or justification.

When you have the members of the team in place, they must work together, trust each other, and not be afraid to make mistakes. *You must want to succeed so badly that you*

---

[10] Why two instead of five? Because most companies will be lucky to identify two.

"In the information age, every-thing turns on the individual. One individual who knows how to use her head is more important than a dozen com-mittees. Isn't that obvious?"

— *Dr. Timothy Leary*
Former Harvard Psychologist in
an interview with **Upside**
magazine, April 1990

not only allow mistakes, but you admit them, examine them, and learn from them. This is a test of maturity, but few firms pass muster. In the infor-mation age you must hon-estly share information to succeed, but many com-panies avoid this.

High Tech ...

You need implementers together with creators

T he problem with needing creative genius is that these people are scarce. If in addition you require them to be efficient and effective implementers, your quest may become impossible. Creative people often don't bother with detail, and it's said that Einstein had difficulty balancing a checkbook. Innovation has two components, invention and implementation.

If you can't implement quickly and well, you will fail as surely as if you had started with the wrong idea. Implementation excellence is required too. You have to get into the market quickly, your product has to have quality, and you must produce it competitively. These are difficult tasks. A skunk works invents well, but implements poorly. I am told that at the famous Lockheed skunk works they did little but test and fly prototypes. When all the bugs were worked out, people were brought in to document the designs and put them into production. This was sometimes hard on test pilots, and it won't work in today's fast moving markets. Even in Lockheed's case, the skunk work's products were custom, expensive, and produced in limited quantity.

You have a prototype, and you know you must get to market quickly. The temptation is to press on with the creative team, since the product is almost done. *In my experience when you have the first prototype, at least 80% of the work remains to be done.* Recall the USS Monitor. It did the job and the enemy's guns could not harm it, but it was a kludge. It had muzzle loaders because there was no time to get modern cannons, the engines filled the ship with smoke, and the turret malfunctioned in the middle of battle and rotated continually. It was low in the water and not very seaworthy. At the time some claimed that it was good to be barely afloat because it made for a difficult target, but the prototype soon sank. Today we would say it had minor technical bugs and was not user friendly.

Unfinished prototypes are the curse of the skunk works, and this brings us to the ultimate "information age" prototype: software. The software industry is notorious for selling prototypes, since it takes tedious evaluation by experts to know what horrors are lurking behind the pretty screens. Today's customers call such pseudo-products "vaporware." High-tech customers abhor vaporware, and they tend to depend on outside expert opinion to judge if products are "real."

Premature release of a vaporware product is usually much more expensive to a company than waiting for the real product to be finished. You want to save money and time by pressing on into production, but you have the wrong team to do this. The team you have is committed, believes in the product, and doesn't want to release it to others who will likely "screw it up." The creators want their intellectual child to be perfect, and will likely "engineer it forever." You have a problem, but selling prototypes — vaporware — is not a good solution. The only solution I know of that works well is to add different talent and different viewpoints to the core team to help finish the product. This is usually not done, because it can be like putting the cats and the rats in the same cage.

You are now entering the "twilight zone" between creation and production. Some companies have formal terms for this phase, when the hardware is more than a prototype but less than a full product. Several units are now being built, but by engineers and technicians. Here is where you make the decisions that may save tens or hundreds of millions of dollars and much grief later. Has the product been thoroughly tested and does it meet specifications with the normal selection of components? Can production units be economically verified and adjusted? Is the unit reliable, and does it stand up to environmental or shipping stress? Can it be serviced? Does it comply with relevant standards, especially safety and electronic emission standards? There are many quality issues that should be addressed *before* you start selling the product.

If your product is software your problems are even worse. Can anyone understand the software, other than the creator? Is it easy to add to or maintain the code? Who has the knowledge and discipline to test your software thoroughly? Who is accountable for bug identification, tracking, and fixing? The norm today is for products to be a mix of hardware and software, and for these you can expect to encounter every possible combination of both sets of issues. These quality assurance items don't place high on the agendas of those in your skunk works, but I can guarantee they are important to others in your business and, eventually, to your customers.

To the engineers, the product is mentally completed and reduced to practice, so obviously any competent person can build it. Good engineers can even show technicians how easy this is, and help them assemble and test a few. The Guru is concerned about delay, since he or she knows the window of opportunity can slam shut. Even the leader will start feeling extreme pressure, since the troops are restless, the burn rate is increasing, and he or she wants to get to market either to make money or to get on with the next project. You need testing, quality assurance, and help from "...ility experts" (e.g., maintainability, testability, reliability, and standards compliance testing), and you ignore this step at your peril.

In the U.S. we usually try to do some type of handoff, and most companies have formal processes for doing this. This sequential approach rarely works well. Naturally, everyone wants to add value, so each time the product is handed off the temptation is to reinvent it, or at least to change it. Some changes are necessary, but the inventor's worst fears can come true as they watch their creation destroyed by implementers who lack understanding. Probably this is better than leaving the device in R&D to be perfected forever, but I don't think it is the best choice. *It's better to add implementers to your creative team.*

We invented it, but our Japanese friends have formalized this alternative.[11] They say, "Stop running a relay race and take up Rugby." In Rugby, the ball gets passed within the team as it moves as a unit up the field. The team retains all the players, all the talent, and the Japanese call the game holistic. They speak of a holistic product development approach with six characteristics: built-in instability, self-organizing project teams, over-lapping development phases, "multilearning," subtle control, and organizational transfer of learning. This is hard to do well in any culture, and it requires exceptional leadership. The concepts deserve some explanation:

1. Instability means that upper management gives the team a broad and ambitious goal, but not a plan. The idea is to allow freedom and provide challenge. A goal example from Fuji-Xerox was to develop a new copier within two years that had half the manufacturing cost and the same performance as their existing high-end copiers. This led to their highly successful FX-3500 product.

2. Self-organizing teams have three characteristics: autonomy, self-transcendence, and cross-fertilization. Autonomy means that management opens its purse, but shuts its mouth. Self-transcendence means that the team goes beyond existing limits, and has the need and the ability to invent new things. Cross-fertilization means that the team is mixed in terms of backgrounds, thought processes, and training.

---

11 "The New New Product Development Game," by Hirotaka Takeuchi and Ikujiro Nonaka, **Harvard Business Review**, January-February 1986, pp 137 to 146.

3. Overlapping development phases is a concept I developed independently and call "the fat arrow." See Figure 3. The old style of project management was sequential, engineering centered, and prolonged. It had the minimum amount of people working on the project at any one time, but took many years to get to market and redirection was common. Typically the design engineers did what they thought was appropriate, and the other organizations were left to sort it out and solve problems. It was common in the 60s and

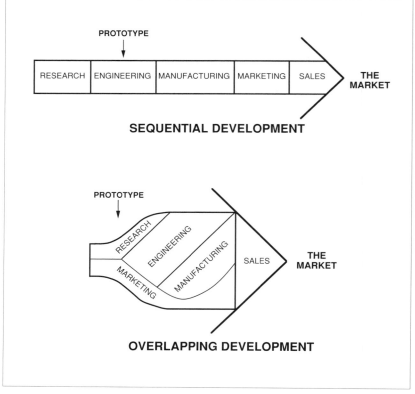

Figure 3. *Project organization.*

70s to succeed while slowly fixing product weaknesses and production problems, but today's competition rarely allows this.

The "fat arrow" puts all the critical skills in place, and optimizes getting to market fast and getting the product right the first time. Over the life of a project, you will invest about the same amount in either a fat-arrow or a skinny-arrow project. The investment to complete either type project — the areas of the arrows — is about the same since the same amount of work needs to be done. The fat arrow spends the money faster, but gets the project done sooner.

The economic return of the fat arrow is *much* better. Some good studies have shown that in today's high-tech environment corporate profits are more affected by product time-to-market than by engineering costs, manufacturing costs, or anything else. Companies that persist in running prolonged projects per the "skinny arrow" also incur a high risk of cancellation or failure in today's markets. Even the best Gurus have problems predicting what product features will be competitive in four or five years.

4. "Multilearning" formalizes the trial and error process. Team members continually test ideas on each other and bring new ones in from outside the team. They learn from competitors, customers, the literature, outsiders, and across functions. You want senior team members to have successful backgrounds in several technical fields and functional areas.

5. Subtle control is almost the converse of traditional management process. You control by selecting the right people, creating an open work environment, encouraging listening to customers, rewarding group performance, and tolerating mistakes.

6. Organizational transfer of knowledge means members should migrate into and out of the skunk works. The composition of the team should change over time to allow the best effectiveness. Junior people should be trained, and heroes should move to key projects to spread knowledge.

    Note that passing down words of wisdom from the past does not work well when the external environment is unstable. In today's environment the best transfer of knowledge is by "osmosis." Moving key individuals and consultants around is better than using formal processes.[12]

The Rugby analogy is not perfect, since the makeup of the high-tech team should change at the endpoints. At the beginning the creators dominate, but at the end the product belongs to the managers and salespeople. The best examples of this cited by the Japanese are U.S. companies such as 3M and HP.

> **Note:** I recently discovered that a (much) larger consulting firm is having success promoting a process-based solution that yields a project structure somewhere between the old "skinny arrow" and the information

---

[12] This is why I say that it may be possible to train 'Gurus" through an apprentice program, but not through a sequence of lessons and examples.

age "fat arrow." They advocate detailed process, a matrix organization, and "core teams." The teams consist of people loaned from functional organizations who report on a dotted line basis to a project facilitator. The project is divided into tasks and activities to be conducted by the functional organizations. The project facilitator collects information, tracks progress, and reports to a "management committee." I suspect this compromise method is less threatening to existing organizations. The existing functions retain control of personnel and budgets and have seats on the committee.

Their approach seems to have three major benefits: 1) The technique works with only ordinary (junior) people on the team. 2) The technique makes management sufficiently comfortable that they will agree to fewer reviews. This saves time, since each formal upper management project review can cost months of delay. 3) The method reduces time-to-market in other ways. In general, the technique is touted based on its ability to consistently and significantly reduce time-to-market through better communications and more limited management meddling.

They said the engineering-centered "skinny arrow" produced worse results than their technique, and claimed that sometimes (where their clients lacked formal processes to coordinate the activities of their various organizations involved in product development) they had cut time-to-market by half. If you still use the skinny arrow or something more primitive, and if independent teams or a skunk works are concepts too revolutionary for your company to accept, you may wish to consider starting with some process-based solution like this. I dislike bureaucracy, but agree this is probably better than sticking with the traditional skinny arrow. It may help, and, if not, at least you will have moved out of your rut.

The role of the implementers is not as obvious as it seems, so I will dwell upon it. Naturally, their first job is to take ownership of the skunk works' idea and move it into the market. If successful that should be a beginning, not the end. Once you are leading a new market, you are exposed to unprecedented opportunity. You know the technology better than anyone, and you have wonderful customer relationships. Your implementers should be allowed to exploit this. Tragically, many U.S. companies stop, disband the team, and want to extract profit. This is exactly the wrong thing to do. Let them do the next product! A successful market entry often creates the opportunity to do three or four cycles of new products. Your implementers can now click like a well-oiled machine, so move them immediately into a related development. You already made the investment and you have the knowledge, so why stop? Usually it will be the second or third product, not the entry product, that is the big winner. The original MAC was a sadly anemic machine. It had limited storage, a mediocre display, poor performance, and lacked a hard disk. Had Apple stopped there the story would not be a happy one, but the follow-on MACs were much better.

Think about this carefully. You won! You took the risk, created a new market, and succeeded in the "suicide square." Now your next products in this market can be low risk extensional or replacement products. You can launch these products into a high-growth market with little competition, since you are the leader. You can do these quickly, incrementally, and with high return. Cherish your skunks for they can lead you into the future, but also cherish your implementers for they can make you wealthy if you let them.

Also remember to keep listening to your markets, since they will mature and shift. To prosper, you must follow — or better, lead — the changes. The day will come when the successful innovation that built your strongest business is no longer appropriate. Many high-tech companies have been stranded as their markets shifted from mainframe computing, to mini-computers, and then to workstations and PCs. The very best companies replace their products with new ones *before* they are forced to. It's a cycle: skunk works, implementation, and production. For a while you can cycle back to the middle (implementation, production; implementation, production), but the day will come when the skunks will save you again.

A sad thing to view in high tech is the team of lonely, isolated, bedraggled innovators.

High Tech ...

Aren't those stupid customers a royal pain?

I was once an officer in a promising new venture. Our products were innovative, and offered breakthrough advantages. We moved into key accounts early, seeking experience, exposure, and revenue. We wanted to secure these accounts before our expected competition, large and reputable companies, got their designs to market.

The strategy worked well, though the products were only prototypes. We sold systems for small trials and test sites, and we won over the market. Orders poured in. We needed to finish the design, and we needed a larger plant. We needed some capital, a few hundred thousand dollars, to expand. We wanted a drink of water, but we got a waterfall. We were acquired by a large company outside high tech, who bought out our original investors and wanted to manage us.

The two cultures clashed badly. I had joined the team late, but I shrewdly had a "no cut" employment contract with assured salary and stock options. In effect, I was sold with the company. The new owners wanted us to "Sell, sell, sell." They wanted a good balance sheet so they could take the company public. I suggested that perhaps we should finish the product first, but this was deemed an excuse for poor sales performance. The creators were asked to leave, since they were considered impractical troublemakers. The new management could walk to the end of the manufacturing line where we had burn-in ovens and see hundreds, then thousands, of units working flawlessly. Well, working almost flawlessly. Money poured in, and the plant and distribution channels were expanded. Inexpensive and obedient junior engineers were hired to testify how good the products were, and they even mostly believed it.

The market was a life safety market, so customers were far more fastidious when moving out of the trial phase and into commercial service. They wanted their systems to work day and night with little or no attention. We could always make the products work some of the time, and with engineers "stroking them" they would even work most of the time. In commercial use, of course, it wasn't good enough, and large systems were hopeless. I wound up as the marketing officer of a company trying to sell a life safety product that did not work. My thoughts were troubled by visions of little old ladies having heart attacks, staggering to our control panel, pushing the emergency button, and receiving no help. This never happened, but I resigned the job and excused the new owners from my contract in return for a few weeks of paid vacation. It proved to be a wise decision.

The company aggressively promoted the product, but they grew increasingly angry with their unreasonable customers. The customers were not only unreasonable, but they were stupid and they complained. This led to finger pointing. The company contended that if the customers took more care and maintained their systems better, the equipment would work properly. The customers contended that the equipment should work properly in the systems they had. Discussions grew hostile, and each side developed technical evidence to prove their positions. Litigation was threatened, and the new President, furious, dared his customers to sue. A few did, most stopped buying, and some would not pay for the equipment they had received. The end result was bizarre. The industry became convinced that this type of equipment could never work well and turned its attention elsewhere. The large companies that were rushing to catch this early leader stopped. The market was stillborn.

The company's doors closed, several years and almost $8M in investment later. The President moved to a new role for the large company, and remembers this assignment with bitterness. He blames the founder for selling "snake oil," and the customers for being "stupid and cheap." Ironically, the original founder still believes in his old vision. He bought the company (and its former major competitor — they had similar problems) for a pittance, and, a decade later and on a shoestring, is attempting to rebuild it. He is reworking years of returned equipment, and maintaining the installed base. He blames the former owners for understanding capital investment, but not people and "soft" issues.

What could have been done to make this example a happier one? I don't have any better suggestions than those I offered at the time. I think patience and a little extra front-end engineering investment could have made a big difference. Think of how our Japanese friends would approach a market like this. They would first invest as necessary to get everything sorted out and develop credibility. Their mindset during this phase would be flexibility and willingness to change. They would value their unhappy customers, because it afforded them the opportunity to learn. After their early customers were happy, then they would promote their products.

Their first project was successful, and they created a beachhead into a new market. They are now far from their company's mainstream, and there is no invasion wave behind them. They are abandoned, and they will get no more help. The lucrative implementation and production cycle is effectively bypassed, and the first product win is expected to bring the company all the way into a happy and prosperous future. Now their companies wish to run the venture like a product division, and they expect it to compete with mature businesses. This is a foolish thing, but it happens often. The Japanese know better than this, so their model is to roll wave after wave of new products into a new market. In the U.S. we seem more inclined to invest to shore up a marginal business than to exploit success, and this is a tragic fault.

There are few things in high tech that can be stated categorically, but here is one. *It's easy for a competent competitor to defeat a single isolated product, however good it may be.* If a competitor positions products below it, above it, and alongside it, and keeps doing new introductions they will eventually prevail. If you allow this to happen, you will have used your best talent to create a market for your competition.

> **"Do not linger in dangerously isolated positions"**
>
> — *Sun Tzu*
> *The Art of War*

Production

With the arrival of the Japanese, so much has been written about manufacturing that I hesitate to add more. In the 1970s U.S. industry was decimated by competition that used such simple strategies as defect-free products, process innovation, and delivery dependability to gain differential advantage.

It's interesting that so few in industry, government, or local communities realize how much U.S. employment depends on electronics production. According to American Electronics Association (AEA) data for 1988, electronics accounted for one out of nine manufacturing jobs in the United States. It employs more people — 2.6 million — than any other manufacturing industry. This is *three times* what the auto industry and *nine times* what the steel industry employs. Also, electronics makes the tools that all manufacturing industries need to be competitive today.

Your skunk works may get you into a new market and build a defensible beachhead, but you need good manufacturing to make money. Remember, it's rare for the entry product to produce a large profit. Usually it's the next one or the one after that makes the most money. It has all the details exactly right, and it offers better quality

> "If you do not have the lowest manufacturing cost, you cannot make your product positioning 'stick.' If you cannot compete at manufacturing, a competitor will eventually take the market that you created away from you."
>
> — *Steven C. Wheelright*
> *Ex-Stanford professor,*
> *now at Harvard*

and lower cost than the entry product. Manufacturing will decide your profits for decades, and it's a game of inches. You have to do many things well, and you have to do them well repeatedly. It was interesting that most manufacturing lines I have toured in Japan had graphs displayed. The lines compared their output, quality, and the number of suggestions for improvement made and adopted. I had the feeling it was like watching the grass grow. If you watched you saw only practiced routine, but if you came back in a few weeks or a month I think a skilled eye would see significant alteration.

> "We value creativity and innovativeness, and we don't value production. But the money is not in invention, it's in production."
>
> — *Michael Dertouzos*
> *Director*
> *MIT Computer Science*
> *Laboratory*

Let me stress this point more extensively, since it is a key one. *No matter how good a job you do at building competitive barriers (with patents, trademarks, trade secrets, distribution channels, or whatever), your advantage will not last forever.* In years or decades, determined competitors will eventually work their way around your protection. There's only so much that the "high cover" of market strategy, positioning, and intellectual property defenses can do for you. Eventually you must either abandon the fight or contest the issue in the jungles or trenches, perhaps hand-to-hand with your adversaries. Long term you have only two choices. You can either abandon the market that you worked so hard to create, or you can fight for it. The better your initial success, the larger the market will be, and the more it will be worth fighting over. If you choose to fight, you win or lose based on your manufacturing prowess.

In DRAMs (basic computer memory chips), most U.S. suppliers — even Intel, a truly formidable company — chose to abandon the market and protect their short-term profits. They had much motivation to make this decision. The playing field was not level. It was proven — too late to help much — that the Japanese were dumping below their costs. One exception is a tiny start-up in Boise, Idaho called Micron Technology. They stuck with DRAMs, they endured, and they developed an excellent manufacturing capability. Today the Japanese are no longer dumping. Based on Micron's manufacturing strengths, IBM has agreed to transfer its advanced technology (4 Mbit DRAM designs) to them. Micron has a tiger by the tail, and is getting one devil of a ride. Their profits are bouncing up and down like a yo-yo, but they're hanging on. They're growing, profitable, and positioned for the future. Unfortunately this was only a minor tactical victory, and the U.S. is losing ground in integrated circuits (ICs). In 1988 the top three IC producers in the world were Japanese, and there were only three U.S. firms in the top ten. In 1978 we led the world: the top four companies were all U.S. and we had six in the top ten.

> "The Japanese economy is growing fast not because Japan keeps us out but because the Japanese are investing like mad and working like hell. America is growing slowly because Americans are doing neither."
>
> — *Robert Reich*
> *Harvard Political Economist*
> *and Writer*

The electronics industry has studied the "DRAM Debacle" extensively, and the facts are now clear. We lost the lucrative computer memory market that we had created to Japan. The reason was not labor costs, since they were higher in Japan. The reason was not excessive middle management, since the U.S. companies were among the leanest and best-managed that we have. One major reason was Japanese dumping, and the large U.S. semiconductor firms contend that this, coupled with Japanese government subsidized technology development, was the main reason.[13] The other major reason was that the Japanese simply had much more efficient production processes and better quality. Many industry observers claim this, not dumping, was the key issue. Cost of capital was a disadvantage too.

A few first rate U.S. companies, most notably Motorola, have taken up the manufacturing challenge with messianic intensity. Top executives are touring the industry, and offering to share knowledge. They caution that there are plateaus where more investment and work does not seem to help, but it's important to persevere. *To compete in manufacturing you must commit to continuous improvement though it may take years of effort and investment to cross the next quanta.*

> **Note:** Motorola thinks this is a very important point, and asked that I emphasize it. If you invest for a few months or even a few years, you will hit a level where further investment does not seem to help. Management

---

[13] Private correspondence with Mr. William Weisz, Motorola's Vice Chairman of the Board. Dr. Gordon Moore of Intel Corporation gives talks that stress this point also.

process as taught in our business schools would suggest that you stop investing at that point. This is exactly the wrong thing to do. If you continue to practice continuous improvement instead, *Kaisen*, you will one day break through into the next plateau of effectiveness.

The Malcolm Baldrige national quality award is a training process and a formal competition, and Motorola requires that all its suppliers participate.[14] U.S. companies can compete with anyone in manufacturing cost and quality if they truly commit to doing so. Motorola is winning — their phones set the standards for Japan — and they just announced a breakthrough cellular phone technology that uses satellites to allow communication anywhere in the world. Xerox endured much pain after Japan finally worked through their layers of strong patent defenses. Xerox also chose to fight for its market. They won and gained the respect of the Japanese in the process. They learned about skunk works, rapid development cycles, competitive benchmarking, and manufacturing. Today they can compete with anyone in the world.

There are many things that can be said about manufacturing, and many qualified consultants and organizations that can help U.S. companies get better at it. Investing in manufacturing does not cost, it pays. Not only

---

14 While most agree that improved quality is important, the idea of expensive quality contests has critics. See "Dubious Achievement," by Patrick Houston, **Business Month**, July 1990.

is it easier to market your products, but the money you save by just-in-time inventory control and by not having to rework defective units more than repays your investment if you allow a long term payback.

It pleases me when TTG's clients include production engineering expertise on their development teams, because the results are usually so much better. My impression is that these examples consistently have better quality, compressed time-to-market, and far fewer problems and surprises. I have also noticed that teams with manufacturing engineers tend to reuse previous designs, and this saves time and money. Most of all, the designs that such teams deliver are *different*. You can look at them and imagine how they could be built and tested economically.

I had a client that staffed its design team heavily with engineers from manufacturing. The product was somewhere between a replacement and an extensional product. It had less than half the cost, one third the weight, and one quarter the volume of the product it was to replace. It used most of the circuit designs of its predecessor, and it was a thing of beauty. It could be completely tested and calibrated on a circuit board

> "There is a very strong tendency on the part of Americans, whether in government or business, that you really have to establish some achievement of your own. In Japan there is great emphasis on continuity. Unless there is something wrong, I build on what my predecessor has built. In the United States the new man comes in and often the value of that man is judged by the things he does differently than his predecessor. Often this is very destabilizing — you start from scratch. In manufacturing, there is a great deal that can be achieved by continuity, the experience curve."
>
> — *Yotaro Kobayasi*
> *President*
> *Fuji-Xerox Corp.*

"flat" by adding just a few jumpers, and it had very few areas where adjustment was required. The prototype was developed by a small skunk works in under a year, and the planned project payback was about 18 months. It would have made a good example for this book, but the project was stopped when 90% completed for unrelated reasons (a facility closing).

All over the industry I hear horror stories of designs that are forced into volume production before they are fully sorted out. Not having seasoned manufacturing engineers on your team saves perhaps 10% on project team staffing, but wastes many times that in opportunity and rework costs later. It also places extreme stress on the people involved, and limits careers. Rational people avoid pain, so few talented engineers seek careers in manufacturing in the U.S. I think this waste and delay would not happen if management took a longer view. This waste is another symptom of budgeting on an annual versus a project basis.

Much progress is required since we are not in very good shape. From autos to electronics, Japanese quality is higher. My personal experience — I do not profess to be a manufacturing expert — is that the exceptional U.S. electronics lines are as good or better than any in the world. Unfortunately these are few, and our norm is far below Japan's. The good news is that we are starting to catch up and learn quality manufacturing. In integrated circuits U.S yields were only 25% in the early 1980s. This means that three out of four silicon wafers were bad and had to be scrapped. We improved our yields to about 75% by 1990, meaning that only one wafer out of four had to be thrown away. Over the same period Japanese yields improved from 75% to 90%. That means the Japanese advantage has decreased from 200% to 20%. It's hard to feel good about this since 20% is still an enormous disadvantage, but we are on the right track. In a few years, as we both approach perfection, our relative disadvantage should continue to lessen further. Manufacturing is not an innate skill. It can be learned, and we are getting better at it.

Even in the auto industry U.S. suppliers are getting better at quality, though not fast enough. Unfortunately the competition has improved too, and perceived relative value and quality are what matters to customers. The "big three" are all still losing market share. Chrysler is a particularly sad case and is reportedly again on the brink of ruin. Lee Iacocca said all those moving words about people, products, quality, and global competition. He said it well, but you actually have to do it to win and Mr. Iacocca became more interested in making deals than making cars.[15] GM made more and bigger deals, and they have lost the most market share. GM bought Hughes and Electronic Data Systems, but I never understood the relevance of these purchases to automobiles. With EDS came Ross Perot, and GM reportedly paid him a princely sum to leave their board because he was so critical of their business decisions. Criticism is painful, but so is losing market share. **Roger and Me** was not a well reviewed movie in Detroit, but it's interesting to watch and is

---

[15] Source: "Myth vs. Manager," cover story, **Business Month**, July 1990.

now available on videotape. In the right environment U.S. workers are excellent at quality. An inexpensive U.S. made Toyota model is in 1990's top 10 for quality, and the rest of the list is expensive luxury brands like Mercedes and Lexus (and, yes, Buick made the list too). For many years Sony's best TV plant was in San Diego, and perhaps it still is their best.

The production of software probably will become "the topic" for the 90s. It's of special interest because so much of high-tech products consists of software, and because software is one of only two technology areas where the U.S. has been holding its share of the world market.[16] Software production has yet to be formalized as a process. Firms like Microsoft Corporation produce their software very well, but formal methods to describe and compare software production processes have yet to be developed. This is one area where U.S. firms seem to be preserving their competitive advantage through trade secrets. It's well known that process advantages usually lead to more sustainable competitive advantage than product advantages. Process advantages are generally much harder for your competitors to learn and copy.

Manufacturing is not the only reason, but we are running a large export deficit in electronics products. In 1988 our electronics trade was about $10 billion negative, but this mostly has to do with Japan since that year's def-

---

[16] A quote from John McPhee, co-author of a U.S. Commerce Department report on the competitiveness of the U.S. electronics industry, **Wall Street Journal**, June 11, 1990, page B4.

icit with Japan *alone* was $20 billion.[17] Our world trade balance in electronics was positive as recently as 1983, it hit a peak of $15 billion negative in 1986 and 1987, but is now improving and even the numbers with Japan are steady. Unfortunately, the reasons for this stability are artificial. In the mid 1980s our electronics trade balance was stabilized by an influx of Japanese screwdriver plants. The parts and assemblies are all imported, but don't count as electronics product imports. The final *products* are tallied as U.S. built, and may show up as U.S. exports.

This is an issue for Europe's 1992 legislation. The Europeans don't like screwdriver plants, and have done exhaustive studies. In one case, the British brought in a Japanese auto plant hoping to help employment. Only 3% of the value was added locally. The net

---

[17] All this is American Electronics Association data.

Is there an absolute for quality?

In the early 1970s a U.S. company decided to contract for a critical electronic component, and after laborious investigations a Japanese supplier was selected. Neither the U.S. company nor the Japanese firm had done business with each other's culture before, so negotiations were a bit confused. The U.S. company insisted on a legal contract specifying the deliverables in detail. One such specification was for quality, and the U.S. company insisted on 99.7%.

When the first shipment of 1000 units was received and unpacked, the U.S. firm noticed that each unit was in a box, but noted that three boxes were of different color from the rest. Perplexed, they called the supplier and asked for an explanation. The Japanese responsible for the account explained that they didn't really understand why their customer wanted 99.7% quality, but they had complied with the contract as agreed. To make it more convenient for their new customer, they shipped the defective units in colored boxes.

impact was local job loss, since the plant's products won market share and this cost jobs at competitor's plants. Anyhow, Europe has served notice the EEC will not tolerate screwdriver plants and are threatening to count U.S. screwdriver plants' output as Japanese.

Important note: I did *not* say that all Japanese plants were screwdriver plants or imply criticism of the Japanese. The creation of these plants is a sincere attempt on their part to ease the concerns of their trading partner. We like to pick on the Japanese, but our trade imbalance with West Germany is just as bad.

High Tech ...

Sales

I was tempted not to include a section on sales. It seems so obvious, and America invented the "Yankee Trader." Many large U.S. companies do well at sales, and those like IBM are the best in the world. Unfortunately, our small to medium sized companies don't do this very well at all, and it costs them dearly. Without the sales function to generate orders, the company lacks money. If a company lacks money, nothing else matters.

What can go wrong with the sales function? Many things can go wrong, and they often do. For starters, many technologists think that if their product is sufficiently wonderful, it will sell itself. This is a version of the better mousetrap theory, and it's just plain not true. It doesn't work that way. Other than possibly your mother and a small circle of friends, customers don't have to spend their money for your products.

The next dumb assumption is that if your product is not selling, all you have to do is drop the price. Your product is far more than the device itself. Perhaps you should raise the price so you can afford more promotion, better product literature, and customer training. I believe in demand curves, but I have seen many cases in high tech where raising the price caused sales volume to increase.

High-tech products are unusual things to sell. For starters it takes a much higher level of commitment for a customer to purchase one. Do you ask for expert opinion before you buy a bottle of soda pop or a brand of tooth-paste? Perhaps you do, but I have never seen anyone willing to admit it. There's little emotional commitment to such purchases, since risk is low. Very few people — only the lunatic fringe — buy a technology-enabled product without thorough evaluation and expert advice. For volume items like PCs they read reviews in the trade press, seek unbiased and experienced users, and look at which brands sell the most. Customers even form user groups to keep the vendors honest. Corporations are even more picky, and for some types of products they may evaluate million dollar demonstration systems for months before they choose a brand. If you screw up and buy the wrong PC, you've wasted several thousand dollars of your money. If you buy the wrong computer system or test system for your company, it may cost you your job or your career. The newer and more novel your product, the harder your sales force must work to move it into the customer's comfort zone.

It's very expensive to set up a distribution channel, and, worse, it takes time. You have to hire salespeople, find them offices, and train them. Many small companies try to use sales representatives. For some products this works well, but there are risks. Sales reps work on commission. There's a saying in the industry, "reps sell what sells." Reps are rational people, and they are independent businesses. If your product takes much time and effort to sell, why should they beat their brains out selling it? You must either give them a logical answer to that question or else make the product easy for them to sell if you expect success. For you to win, they must win too, and you should acknowledge this if you distribute through reps.

One of TTG's clients had a strange situation recently. A distribution company sought them out for a new type of product, and negotiated an exclusive distribution contract and a good sales commission. The distribution company's sales people were top notch. Some were crossovers from real estate sales, where they had been the highest producing salesmen in their entire companies. They were doing well selling some electronic products, and had access to the high-level executives in the targeted industry. After launching the product several months passed with no orders at all, but during the same time my client's technologists had sold some units with little effort. Upon investigation, it turned out that the sales people were leaving these new products with their high-level manager friends to evaluate. When they returned, guess what? There were no orders, and nothing had been done with the demonstration units. The busy managers had done nothing with the unfamiliar product, but they were embarrassed and, so, raised naive objections about it. I think this was a new experience for my client's sales channel. It was equivalent to showing a house to a prospect, and when they tried for the close having the prospect say "What is a house?"

For a new and innovative product, the early order levels can be directly related to the number of actual sales demos given to those whom you expect to use the product to do their jobs. You must get into the trenches with early customers to create demand, and this is tedious. You can't make ten sales calls a day, and it's hard work. You have to show your new prospects, not tell them, the value of the product and how to use it. You probably must give repeated demonstrations to the same customers, and you will certainly have to help the actual end users develop comfort with the product. (Can you imagine having to show a customer how to use a house or a bottle of soda pop? In high-tech markets, you often must show and instruct your customers.) Initially the demo-to-sales conversion rate is very low. For each, say, ten or twenty demos given, a good salesman might initially get one order. As the product becomes more familiar, this success rate will increase dramatically. After the products are endorsed and accepted, repeat orders will eventually come with no demos at all.

Fortunately, my client's situation is self-correcting since the contract I helped them negotiate granted exclusivity of distribution based on *monthly* order performance. The client's engineers have now stopped selling the product, and the distribution organization has had the situation explained to them. They have an unusually generous sales commission structure, and they are reportedly now out in the trenches. They are still making rookie mistakes, but I think they are trying to sell my client's products. I hope they succeed, but even if they fail other good sales reps want the products and are waiting. As they say, "You pays your money, and you takes your chances."

There are much worse situations. This product costs only a few thousand dollars, and has a purchase cycle of a month or two. Some complex products with higher levels of commitment may take a year or more to sell to key accounts. For this, you need a sales salary base, plus commissions.

Why not just make some money?

In early 1989 TTG was retained to help a small Oregon client called TriSys, Inc. The company was a new venture formed to do specialized testing of CMOS integrated circuits. Their products are innovative computer-based systems. They take ordinary items like 386 PCs and oscilloscopes, but they add their unique (patented) electronic sub-systems and software to inject precisely controlled high-voltage electrical pulses into each pin of an integrated circuit. This is called electrostatic discharge (ESD) testing.

The company was growing, so they hired me to help them decide the next product to do. After a few months of study it became apparent that 1) they were too small to finance diversification, 2) they hadn't even scratched the surface of the revenue that could be produced for the products they had, 3) the existing products needed extension, and 4) if they did not quickly build a good distribution channel their future was not very hopeful.

Comment: To pursue new ventures or new products seriously, you must commit *both* marketing and technical talent. If you cannot do so, it's probably better to wait.

My assignment for TriSys soon turned into a "chicken-and-egg" problem, since if they had a good sales channel they should have more products to sell to generate revenue to support the salesmen. Fortunately, when you have a unique product, good distribution, and good customer access, it's not too hard to seek out complimentary products.

I induced the company to hire a full time V.P. of Sales, and helped them select a qualified candidate. The candidate was hired, Mr. Wyatt Starnes, and he did well in the job. The company was sold, and the CEO, my employer, accomplished his financial objectives and moved on, first to become the board chairman, and then to other things. Wyatt is now the President and CEO of the company, and is reportedly doing well in the job. We never did the next product, but the assignment was productive and the client got fair value for their money. I finished my assignment for TriSys, but the former CEO is now a friend and fellow consultant, and we are doing some work together.

Good salespeople are like a coin-operated machine, and I think a lucrative commission scheme should be a part of every sales compensation package. I even think the commission rate paid should go up as performance increases, but this is heresy in most circles. Most companies place caps on their sales compensation, but I have never understood how this benefits them. I have heard it said that it would be terrible if a salesman made more money than the top corporate executives. I don't know why this would be so bad if the salesman performed exceptionally and brought in enormous revenue. Isn't that what you *want* to happen?

If you play in the suicide square, you will run your sales force very hard. During the entry phase, you must gain market share while your high cover of product positioning and patent protection is still in place. At this phase, the competition will usually try to FUD you to delay your orders until they can develop competitive products. Your sales force must blast through these tactics, and rapidly convert your potential to orders, revenue, and market share. Typically, you will gain *all* your market position in this first phase, so you should move fast.

In the "end game" phase, your competition will have arrived. If you're in a large market and you didn't choose to sell your venture to a large but late competitor, now your salespeople must sell toe-to-toe against the likes of an HP or an IBM. You must fight to keep what you've gained, and you should do this profitably since your competition has deep pockets and prices will be coming down fast.

To play the end game you need top quality, low manufacturing cost, and very good sales. The workstation market is moving into this phase in the early 1990s, and one observer has likened it to the battle of Waterloo. Sun, HP, and DEC have fought each other to a standstill, and now there are drums in the distance. Just over the horizon IBM is on the move with limitless resource, fresh troops, and, finally, some competitive products. Simultaneously companies like Sony are trying to clone Sun's products and drop prices aggressively, and companies like Apple are ranging into workstations because they think the PC markets are too competitive. This is an interesting battle to watch, but not one I would care to fight without major resource behind me. (TTG saved a small client money and grief just by convincing their President to avoid this market area. Their potential hardware platform partner collapsed into serious financial distress a few months later.)

There's another thing to keep in mind about your sales force, and it's an important point. Your salesperson is your corporate face to the customer, and your on-site presence. One of the most flattering things you can do for customers is to involve them in your new product planning. This is good business practice, but it's a serious mistake to bypass your sales force. Over time your business success depends on your relationships with customers, and your local salesperson must play a key role in these. The best practice is to visit your key or leading-edge accounts with your local sales person, a technologist, and a marketing Guru. Conducting good customer-requirements interviews is an art, so you need your Guru present to guide these. It's the content of these interviews that matters, *not* the process. The questionnaire techniques used for consumer products rarely help, and may do harm. There are a few basic points worth mentioning. You should ask questions, listen attentively and sensitively, and take detailed notes. If you choose to serve this customer, periodically return to share your progress with them. If you don't choose to serve them, stop wasting their time.

Each customer visit by your plant personnel is a serious business investment, and it should be managed as such. When the team returns, their notes should be reviewed carefully, and then documented. Your marketing Guru will find this information useful as he or she searches for insight and patterns, and some small portion of the data collected will literally be worth its weight in gold. ("Which portion?" you ask. To answer that you must spend money. The answer is subtle and context specific, and it takes time, expertise, and effort to know.) If you work closely with your customers and your sales force, it will help make your future products successful and help sell more of your current products.

High Tech ...

Are we missing something?

I n researching this book, I sorted through many high-tech writings and inevitably found the literature full of information and examples about the Japanese. Strangely, something seems absent.

There are libraries full of information about Japanese manufacturing, about their quality, about their inventory control, and about their culture. The difference in their capital costs (a three to one or better advantage) is documented and tracked. The loss of markets and the trade imbalance are commented on endlessly. Finally, we now see much written about how the Japanese are getting very good in technology, and that they are buying Western technology. It's amazing to me how disparate our perceptions are of relative strengths.

> "Opportunities multiply as they are seized."
>
> — *Sun Tzu*
> *The Art of War*

The data says that 63% of Americans think that Japan is stronger in science and technology. Paradoxically, 63% of the Japanese are convinced that America is stronger.[18] I make my living in technology and I'm convinced that the Japanese perception is correct. A recent compilation of twelve key technology areas showed that we were ahead or at parity

---

[18] Source: **Wall Street Journal** cover article, June 25, 1990. They quoted such sources as NBC News, Nippon News, and Wall Street Journal.

with the Japanese in all, but our position in new products that exploited these technologies was much weaker.[19] We're (still) stronger at technology than they are, but we don't know it. We're losing markets to them, but we attribute this loss to the wrong reasons.

> "I felt they may well come up with something new that was even more important than the transistor . . . [but] there is not one person there who is thinking about how to use the new technology they are developing as a business. I think this is one area where the U.S. comes up wanting."
> — *Akio Morita recalling a tour of Bell Labs*

A typical story of the difference in philosophy between the U.S. and Japan is the tale of speech recognition. In early 1990 both MIT in the U.S. and NTT in Japan had working prototypes of speech recognition machines. Each device can understand and react to human speech, but ours is better. MIT's machine can understand about 400 words, versus NTT's limit of 32 words. Unfortunately, ours is a research project and there are no current plans to commercialize it, while theirs probably will be included in a product next year.

Interestingly, I could find little written about what the Japanese perceive to be one of their major strengths and a weakness of ours — marketing. They don't conceal that they are market centered, but it seldom is written about here. It comes up, if at all, as a concern the Japanese have when trying to get better at research and technology. In their planning meetings, Japanese technologists complain that they can't get research budgets unless markets are identified. They note that their

---

[19] **Business Week**, June 15, 1990.

corporate R&D has been market driven, and they debate the need to shift to a better balancing act between "market needs and technology seeds." Some want to move to more long-term laboratory-based innovation.

Japanese executives are proud of being market driven, and say their traditional strength is filling market niches quickly. They speak of "Sharp shock" and "Casio shock" to describe the waves of high-quality low-cost calculator products that blitzed the market and drove 48 Japanese companies out of the business. Their model is to find a need, apply technology to fill the need, and then roll wave after wave of products into the market. It's like watching a steam drill slam relentlessly through granite rock: *Study, Target, Niche, boom, Niche, Boom, Niche, BOOM!* This is the essence of market targeting, and companies that can do it earn the respect of the Japanese. In 1982 Honda unleashed some 30 new motorcycle models in a six-month period when openly challenged by Yamaha. The President of Yamaha didn't commit ritual suicide, but he did publicly apologize, concede the market, and resign.

> "One could argue that our most significant miss has been in marketing — something we in America pioneered as a management concept nearly thirty years ago, but of late haven't practiced too well."
>
> — *John F. Welch, Jr.*
> *Chairman and CEO*
> *General Electric Company*
> *(October 28, 1981)*

I've heard U.S. high-tech executives talk endlessly about the Japanese, but I don't remember Japan's marketing prowess coming up. We worry about everything else, but I can't recall discussions or conferences where someone said, "We should worry about how well the Japanese do market targeting." Conversely, when I explain what TTG does (several things, but we center on fitting technology to needs) to Japanese executives, they nod knowingly and say things like, "Oh yes, that would be good. Your companies do not do that very well. Do you make much money?"

The Japanese are *excellent* at market targeting and market creation. Isn't this obvious? Why is this prowess not discussed in technology or business circles? Once they drive an entry into the market, they are diligent at expanding their advantage. We don't seem to notice this, or, if we do, we don't want to talk about it. This gap in our awareness seems strange. We in the U.S. are preoccupied by what Japan is doing, but we seem blind to the reasons that we are losing our markets.

Note: Perhaps this has to do with how we teach business and marketing in our colleges and graduate schools today. You will notice a lack of intersection between the contents of this book and the syllabus of a graduate level business or marketing degree at most U.S. universities.

High Tech ...

Ventures versus adventures

T he VCs say, "A little better is danger- ous," and they are correct. You need a significant differential advantage, and you must move the market far enough that existing competitors cannot reposition their products to block your entry. The customers you want for your new product are already buying something else, and those selling it have more money and better cash flow than you do. They can cut price or raise promotion level with the stroke of a pen. If you don't offer clear advantage, they will FUD you aggressively. Small companies most often fail because they don't move far enough.

Strangely, big companies fail most often for the opposite reason. They want home runs, and they want them with the bases loaded. They want single new product lines that can produce ten or twenty percent of their cor- porate sales in a few years, and this can make the hurdles hundreds of millions of dollars in revenue. New markets start from zero, by def- inition, so the marketers have severe prob- lems. To find a large market they either need to displace a well accepted existing method, or rip an emerging market away from those who are creating it. The technologists have severe problems too, and they either wind up working on an extremely risky technology or needing several risky technologies developed in parallel to work together. Even if the team can convince management to make massive investments quickly, the odds of getting this to come together are not good. Even worse is that the large markets you want to attack are usually populated with formidable predators that have claws, teeth, speed, and endurance.

You need to move far enough that competitors cannot easily react, but not so far that your task becomes impossible. Perhaps the best help I can offer is several stereotypical examples of success and failure described in the accompanying sidebars.

Keep a portfolio

The problem with totally new products is that it's not always possible to predict which ones will succeed. Just as even a good base-

---

Isn't strength and market position an advantage?

In the early 1970s I managed a small product group for Tektronix (Tek). HP owned the market for testing cables on military ships, aircraft, and weapons platforms. Their product was a "sweeper cart," a large $60,000 system. Both HP and Tek sold laboratory time domain reflectometers (TDRs), a type of pulse radar that plugged into their then current lines of oscilloscopes for use by engineers. At the time both companies had comparable technology, manufacturing costs, and financial resource, but HP had a better sales channel to the military market.

We entered the market against HP by offering a simple test set based on TDR technology. The unit was small, battery operated, rugged, and could be carried around and operated by a single, low-skill technician. It sold for about $3,000. Critics pointed out that HP could easily copy our product. This was true, but for years it was Tek's fastest growing, highest margin product line. We had moved far enough to displace the market. HP could react, but not with their present product or anything like it. To copy or improve our test set would take them several years. This was a small market, and one not of strategic interest to them. A perfect product would only convert their $60,000 system orders to $3,000 orders sooner. Since they were a competent competitor, they invested elsewhere and did well. We did well too.

---

ball player makes an out most times at bat, so even the experts often miss in high technology. You must not only discover, but often create, markets for new products, and this is

---

Why not go for the gusto?

Perhaps the best recent public example of adventure was Tektronix' foray into Computer Aided Engineering (CAE).[20] In the early 1980s Tek grew concerned that engineers were moving from test-based design to CAE. A small group wanted the company to pursue this, but didn't secure support. It was not possible to prove a large market demand, because the market was just emerging. The CAE opportunity was obvious to some, but speculative and unprovable to others. The advocates left in frustration, and started a company called Mentor Graphics.

In a few years there was market data and it showed that Mentor, now a sizable company, was a leader. Tek's management grew frustrated as they watched their customers' engineering purchases shift from oscilloscopes to computer-based tools. They started a new project, but it was much too small to compete in what was now becoming a large and fast growing market. To speed progress, Tek bought a small ($6M in sales) Silicon Valley CAE company. They wanted a presence in CAE badly, so they gladly paid $75M and announced to the world that they would be a major player.

Unfortunately, the product proved to be vaporware. Tek threw good managers and engineers at it, but the task was to fix software consisting of a million lines of poorly documented and patched together code while supporting existing customers in a fast changing market. Important customers were depending on the products, so corporate prestige was at stake and exit costs were high. After several years and an additional $200M in investment, the division was sold (ironically, to Mentor) for $5M. Tek even donated a promising new venture into computer aided software engineering (CASE) as part of the package. CASE is now a Mentor division, and is doing well.

---

[20] This example was widely reported in the trade press. For example, see **IEEE Spectrum**, "Tektronix: no more Mr. Nice Guy," August 1989, pp 41-45.

tricky. You will fail much of the time, so why even bother?

Many companies don't bother, and individuals profit from it. Until the end, you can invariably improve today by neglecting the future. It's easy to improve profit over the short term. Simply layoff your engineers, fire all in marketing but your salespeople, sell the buildings, and leverage your assets. Then you can take the money and run before the future comes crashing down on your company. This book is not written for those so inclined, but the strategy is popular in some circles.

To prosper over the long term, you must invest in the future. The day will come when

---

How much technology is too much?

A new venture was formed jointly by two large electronic companies. One was U.S. and one was foreign, but both were world leaders. One led in chips, and the other in electronics systems. They had proprietary technology, exceptional technologists, much money, and good distribution channels. This appeared to be a marriage made in heaven, so the two partners set out to develop a new market for parallel processing computing. They targeted a high value-added area, mission critical computing, and the target markets were Wall Street and the U.S. Department of Defense. The competition, companies like IBM and Tandem, used prosaic technology, but they implemented it well and had good customer relationships.

After a short period of beta testing it became apparent that the losses were increasing on an exponential curve. The market chosen needed perfection, but new technology is rarely perfect. Schedules slipped, and the technologists drowned in an endless sea of implementation detail. The chips were new, custom, and with problems. The architecture was in similar shape, as was the operating system, the compilers, and the applications. As with creation everywhere, it was a technologist's dream, but a business person's nightmare. There was just too much to be done, and unless all components worked perfectly there could be no product. Customers demanded total confidence in fault tolerance and system recovery. More testing revealed more

---

you should move into a new market area, either to continue growth or avoid decline. When this happens, you can either create a market or capture one. If you can find an attractive market that has incompetent competitors, by all means go for it. Simply reverse engineer the products, improve them incrementally, and take the market. Our Japanese friends have done that for years, and they won almost every time by taking advantage of their superior manufacturing processes and willingness to take lower margins. They know most western firms would sooner die than give up next quarter's earnings.

---

problems, which caused further slips. Confidence eroded, and the competitors eventually FUDed the venture into oblivion. When corrective action failed, the partners abruptly cashed out.

The lesson here is that picking the wrong market entry point may make the task impossible no matter how much talent and money you have. The device was excellent, and this venture had phenomenal engineering. They even had a Guru, but they did not add him to the team until long after the business direction and product features were selected, and that was much too late. From a distance, it appeared that this venture lived and died as an interesting and challenging technology creation exercise. It never developed a realistic business focus.

Note: I've been asked, "What would have been a preferable market entry point?" It doesn't matter now since the venture is dead, but almost any place further from their weaknesses and their adversaries' main strengths would have been better. If you had a product with excellent communications hardware and software, disk mirroring, and superior fault recovery, imagine where else you might position it. I can think of several possible choices, starting with file servers for large corporate LANs and ending with better distributed processing systems for air traffic control.

This strategy doesn't work against first-rate competition, because they can make it far too costly for you. With a comparable product, you can estimate a market entry cost of 70% of the leading competitor's annual sales.[21] Even worse is that such toe-to-toe share battles lead to price wars, and this can make the entire industry's profit margins unattractive.[22] In such a case TTG would prefer to

---

[21] Source: Davidow's book. See bibliography.

[22] If you wish to ponder the impact of increased competitive rivalry, Porter's book makes good reading. See bibliography.

Where can a small skunk find shelter?

In the late 1980s I was the business development manager for Tektronix' corporate research laboratory. We had invented a clever, real-time, parallel processing scheme for digital signal processing (DSP). A division was interested, since they had for many years wished to enter the market for real-time analyzers. It was a large market, but it was populated by formidable competitors including the Japanese and HP. A problem was that Tek was doing poorly and downsizing continually, so it was hard to get management to make and keep funding promises.

We formed a new product team, and I was assigned the business development role. We looked at the market. The largest market by far was low performance, under 100 KHz. There were a few higher performance products at higher prices. These were older products, but still selling. They had a scattering of applications, and we stumbled across an area where these products were bought and used though they were not adequate, because nothing else existed. Our technology could be moved up in performance to meet these needs, which came from the signal analysis community. The customers were performance demanding, and they understood technology and would work with us. If we positioned our product here we would have good margins, it would take a major development cycle for competition to react, and we had good patents to slow them down. We decided to create a new market, and were careful to do so in a way that excluded the existing products.

represent the defender. Occasionally frontal assaults carry the day, but the battles are bloody, prolonged, and very expensive.

A more hopeful strategy is to create a new market where you enjoy an advantage. This way you can outflank the established competition. Though you can't tell if any one product or company will win, the economics of this approach can be good. The trick is to invest in the right portfolio over a long period. The portfolio concept is the idea that built the venture capital (VC) industry, and risk was actually quite low.

We asked the key customers if they would share the development risk with us by ordering early prototypes and evaluating them. They agreed, and, with commitments in hand, we persuaded our management to fund the program. The customer's involvement changed our performance targets and features, but we carefully kept the product general purpose and commercial. We called the product Shiva because it had many processors doing work together, and this gave it identity and a soul. We did Shiva as a new venture, and with a small, quick, team. When announced, Tek called it the 3052. Shiva was priced at five times the competition, it won the most innovative instrument of the year award at WESCON 1987, and it created a new and profitable market.

Comment: Shiva was planned as a market-entry product, and it succeeded as such. The original plan was to use this as a beachhead to move into larger markets, but corporate turbulence prevented this. The beachhead was won, but that was the end.

Remember the Japanese steam drill: *Study, Target, Niche, boom, Niche, Boom, Niche, BOOM!* It did not happen this time. Three years have passed since Shiva was introduced without a significant follow-up. Shiva is instead being squeezed for profit. This is commonplace in the U.S. and some think it's good.

How the early VCs did it

I should explain the concept of a venture portfolio. A typical VC fund works by investors putting together a limited partnership. The limited partners invest money and hope to get a better rate of return than they could elsewhere. The general partners sign up enough investors to reach critical mass, and then they look for businesses in which to invest. In theory, the general partners are experts at selecting investments in the targeted areas, so they are paid fees for managing the fund. After a period, the fund is "liquidated" and the companies are sold. Typically the VC fund has several tens of millions of dollars to invest, and the typical criterion is to select companies that promise ten times or more return on investment over about five years. If you want the VCs to invest in your company, they will put up the money you need in return for a major share of your company. VCs want to "become liquid," to cash out and take profits, so their strategy requires extreme growth and a financial market that covets technology companies. I use the words "professionally managed" to describe funds with over $15M in capital, and with at least one seasoned partner.

Through 1978-1986, there were over 1000 professionally managed VC funds. Only two or three failed, and the average annual return was over 28%. (Before you leap to call your broker, I must remind you those days are long gone.) Patience was necessary, even then. The average successful venture took over seven years to break-even, and over eleven to pay back the investment. Typically the losers became apparent after two or three years and were weeded out, but the big winners took more time to become obvious. *There were no useful short term ways to predict performance, so it was necessary to wait and watch.* One third of the ventures failed, but enough paid good returns to produce these numbers. Only 10% delivered the desired "better than ten times" return, but a few of these did much better, and another 8% did between 5 and 10 times return.[23] See Figure 4.

The key concept is that by investing in a portfolio of technology ventures, you could spread the risk in a way that makes your overall investment safe. In the late 1970s and early 1980s this worked very well, and the VCs made very good returns even though their success rate was only 10%.

---

[23] Source: TTG research.

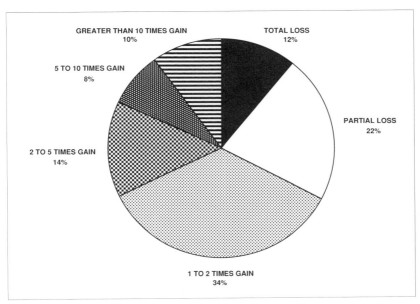

Figure 4. *Rates of return for 200+ VC-backed ventures (1973-83).*

What went wrong?

More recent VC statistics are hard to come by, but everyone knows that VC performance is much worse today. When I mention the old performance numbers to people in the industry, they give sad smiles and shake their heads. A friend closely connected with a major VC fund said, "Today's numbers are not good. You can do better putting your money in certificates of deposit." My guess is that about two thirds of today's investments are in trouble, and that the big winners are down to a few percent. This is a dismal record, but it doesn't have to be that way.

Note: When challenged on this point in the March 7, 1990 issue of **Forbes** Hambrecht and Quist, the well known Silicon Valley VCs, released some internal data. Their funds' performance, except for the 1981-84 period of formation, was good.

In all fairness I should point out that there is lag in the system, so what is reported today depends on the decisions that were made five or ten years ago. There are smart and capable VCs, and some do better than others. It's possible that today's VC investments are better, though we won't know for sure until the mid 90s.

For practical purposes the fortunes of the VC industry may not matter to most technology companies today. The first round of financing is called "seed," since this is where companies start. In all of high technology (hardware, software, and bio-tech) the VCs funded only 33 "seeds" in 1989, down from 41 in 1988.[24] This is a trivial amount of money.

Why is it so small? Some say because it's more fun to play golf, and safer to fund later, larger deals. Only the stronger funds survived the 1983 crash, and few have been formed since. Most partners are, therefore, getting up in years and making good incomes. Most funds today are prestigious, playing with institutional money, and have reputations to protect. It takes much work to judge the merit of a brand new company with no history. If you were a partner in a respected Venture Capital firm managing several $50M funds, would you want to spend weeks of your life away from your business in remote areas (like Oregon) evaluating highly problematical and imponderable issues to protect only a few hundred thousand dollars of one of your fund's money? Conversely, would you trust the valuation of a junior associate when your firm's reputation was at stake? The risk is highest when you are the first to fund a new company, and if a VC invested in one that failed (many will fail) the entire industry would know it. His peers would laugh and he (and his firm) might suffer damage *even if the amount invested was small*. With less work and less business risk, he can invest more of his fund's money at a later stage. Better yet he would then be investing with the herd, and several funds would share the risk and the research. A bigger problem is that the U.S. VCs are not financing many high-tech companies in the late stage "big money" rounds either, because foreign capital is so much cheaper.[25]

---

[24] Source: Technologic Partners, **Computer Letter**, vol 6, #19.

[25] See the "Tokyo Connection," **Inc.**, February 1990.

It's useful to ponder why the success rate of technology ventures, never good, got so much worse in the 80s. It should be obvious that the attributes for success are in scarce supply. U.S. money is more expensive and impatient than capital in any other major industrialized country. A long-term investment in the U.S. is measured in days, and the tenure of a CEO in quarters. If you expect to harvest profit from a technology-based new venture in a few years, you will be sorely disappointed. Usually a venture portfolio value drops for the first four or five years, so looking at the numbers doesn't help much. You have to understand your market and technology and know what you are doing, but much of the experienced VC talent got rich and retired. Their successors and associates sometimes substituted "MBA process" for knowledge.[26] Since few new funds were formed in the 80s, little new blood and new ideas entered the field. The markets changed, niched, and got tougher to understand. Communications are rapid, and competition is world wide. Product life cycles shortened, new technologies emerged, it took more money to seed ventures and much more to launch them. Most important, there was more money available than talent. I actually heard a presentation in the early 1980s that claimed the time to start a venture was not when opportunity was ripe, but when funding was easily available.

---

[26] For an amusing article on this see the cover story "Has Silicon Valley Gone Pussy?" in the June 1990 **Upside**. (**Upside** is in many ways the high-tech equivalent to **People** magazine. It's just now starting to acknowledge that life exists outside of "the valley.")

My impression is that by 1983 the VCs were desperately throwing money (they had a lot of it, and their job was to invest) at anything that was warm and moving. See Figure 5. The investment community does not admit to being emotional or following fads, but I think they are and they do. The financial community discovered high tech in the early 80s, and it became fashionable. Annual investment increased from a few hundreds of millions in the late 1970s, to almost 16 billion dollars in 1983. A literal flood of money was available, but there were only a few companies seeking investment. The VCs wound up investing in things like the 27th PC clone, or the 15th LAN, or the 12th optoelectronic test company. The outcome was predictable. When returns

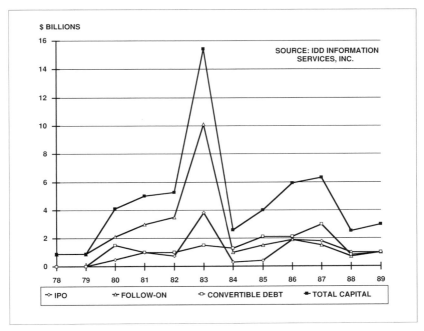

Figure 5. *Total capital raised for technology companies (1978-89).*

started to sag, the hurdles were moved higher, so the success rate went down further. Some VCs lost their shirts, and many smart and seasoned entrepreneurs avoided VC funding because the terms and conditions were unappealing. Today there is less money seeking investment in high technology, and electronics and software are definitely out of style. The lesson is worth pondering, since these were not stupid people. *Even good people with lavish funding will fail in high tech if they pursue the wrong products. If the hurdles are set too high, failure becomes common and the smartest may decline the race.*

Managing your risk

The trick is to develop a good portfolio, understand the business (the market and the technology), and have a long-term view. The problem is, how do you know what to put in your portfolio? I bet on people (track records), technologies, and I want to spread my investments around the playing field (refer to Figure 1).

If I were King, I would want some investments that are safe and produce short term cash flow, and replacement products do this nicely. These products will put bread on my table and pay my bills. I also would want some that offer growth, and these are the extensional products and ventures. These endeavors have more "sizzle." They will let me enjoy better profit margins, and allow me to attract better talent and more financial support. Succeeding here will not only make money, but it gives me a chance to develop closer customer relationships. It may allow me to sell more of my core products. Now I don't have to eat bread alone, I can occasionally enjoy steak. If I could just compete well in these sectors, I would be doing better than 80% of the industry. But in my Kingdom, I want leadership, fame, and wealth. I fear that in today's competitive world if a company does not aspire toward the stars, it will soon sink into the mud.[27] Therefore, I would move a significant portion of my investments into totally new technology and market areas, and I would choose to compete in the "suicide square." One major win here will produce fortune beyond my wildest dreams. If I could win, I could have some champagne with my steak and I could attract the best talent in the world to work in my company. Everyone who works in technology to some extent shares these aspirations, but few know how to invest in the right manner to assure a high probability of achieving these results.

Figure 6 is helpful. Management should protect investors from risk, and most companies do it per the vertical box. They stack up their projects, and the replacement products are well known and promise large returns. The new products are very uncertain, and the extensional ones somewhere in the middle. The arrow represents the "cut line," which moves up and down as the fortunes of the company wax and wane. Unfortunately, this makes it hard to get to the new products, and even the extensional products tend to be turned on and off seasonally. A better way to manage resources in high tech is per the

---

[27] If you are not willing to take risks and invest for the future, high tech is probably not a good career choice.

right, horizontal, box. I'm told that Hewlett Packard (HP), a fine company, does this, not by edict but by practice, and the Japanese are noted for consistent investment in the future too.

The rule of thumb at HP is 20% new, 40% replacement, and 40% extensional, but the numbers matter less than the policy. One point that should be made emphatically, is that doing new things should never be an excuse to ignore present things. To succeed in high tech you must do both, and well. In the

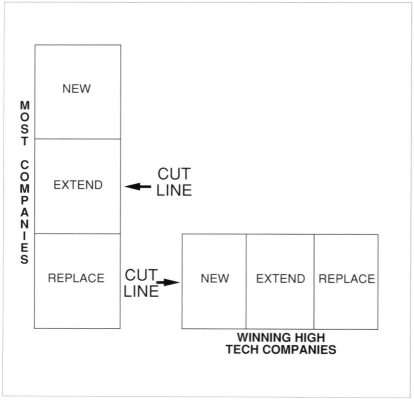

Figure 6. *Project prioritization.*

mix described above, 80% of the company's investment portfolio is improving and extending their present products. Such investments are very profitable. The Japanese have made continuous improvement, called *Kaisen*, almost an art form. Smart workers are your main asset, and part of everyone's job should be figuring out how to do their jobs better.

Comment: Some may not be interested in small $10 to $50 million dollar businesses, but I still suggest a portfolio of ventures is best. Everyone knows that Compaq hit $100M in one year and that it had never been done before. Less known is that Compaq also burned $30M in equity capital the first year, and $32M the next. Few companies have the connections to raise such levels of funding when they are just starting up.

If you have $30M to invest, you can look for the next Compaq. Alternatively, you can start a sizable portfolio of high-tech ventures. If you do the latter and just match the old VC numbers per the fundamentals outlined, you should enjoy a greater than 25% annual return, against a less than 1% risk of losing your investment. A premier team can beat these numbers by a considerable amount.

"But," you say, "We are neither Venture Capitalists nor a large Fortune 500 firm. We cannot afford a portfolio." Sure you can. One thing we have learned in Oregon is how to do new ventures on a shoestring. The total amount of venture money available here is only a few million dollars per year. That's not enough to throw a good party in California, but it's enough to do some venturing.

If you're smaller than, say, $5M per year, and working in high tech you *are* a new venture. You must bet totally on what you are doing, and your core business should be defensible and in an attractive market area. If it's not, usually you had best change something and fast. Companies over $50M per year can easily afford a small portfolio, and those over $100M probably need one to preserve their vitality.

The in-between companies, those between $5M and $50M, should consider spin-offs, partnerships, or joint ventures. You may lack the funding to launch a major venture yourselves, but you possess decisiveness and knowledge that you can profitably leverage. To get to a prototype and a patent position in a new area can be done with low investment, typically only a few hundred thousand dollars of seed capital. If you solicit investors before this point you will pay dearly for this "high risk money." Typically, you must give up 30 percent or more of the company just to get this small sum for your prototype and early customer research. This part you should fund yourselves. A typical high-tech venture costs perhaps a total of $5-10M to launch and grow. With a first-rate team, a good prototype, patents, customer contact, and time, you can raise several million dollars for perhaps 30 percent of your company. Those terms are more appealing, and this money gets you to a producable unit. Then you need more money, but your bargaining position is good if you have prepared properly. For the last kick, you need perhaps another $5M for promotion and production. This low-risk money should cost you less than 20 percent, and perhaps you can even use debt. If you take foreign money you will give up less, and many worthwhile ventures can be done for less than these sums.

Even in today's poor capital markets, a competent high-tech company can spin off ventures that grow to $40-$50M in sales over six or seven years by investing less than $0.5M of their funds at the seed stage. With the scenario above, the parent company could retain about 30%, the venture's management team about 20%, and the funding partners the rest. At a sales level of $40M, the valuation of your venture is about $80M. If you win, you get a 48 times return on your investment.

Pushing the limits

How can you get the money to start? The typical high-tech company spends at least 40% of sales on R&D, marketing, and administration, say, $4M for a $10M company. If you divert $0.5M (12.5%) of this, you can spin off one new venture each year. This offers a career path for your best people, and if you don't have a line of your best people waiting for this opportunity something is wrong with how you are running your company. Just make sure that they have a replacement trained for themselves before they're allowed to move. If you staff it properly, other than the initial formation and board meetings the venture won't consume much of your time. In fact you must ensure it doesn't drain your time, because you have your core business to run and grow. That's a reason to give your team

Why is everything so big in Texas?

Compaq is arguably the most successful new venture in history. They did $100M the first year, $200M the second, $329M the third, $500M the fourth, and $700M in the fifth. During the same time they grew from three employees to over 2,000. They are one of only two companies able to *consistently* penetrate the primary distribution channel of PCs, the dealer, and able to *consistently* penetrate the business market. The other is IBM, a 70 year old company with $50 billion or so in annual sales.

They run their business based on three key precepts. First, the importance of marketing is equal to or greater than the importance of the technology that goes into the products themselves. Compaq's product developments are driven by identified market needs. Second, having the proper philosophy and culture is critical in being successful and achieving high growth. The process of selecting the right people and then making certain that they are doing the right things forms the foundation of their company. The third precept is to manage the scope of the business plan against capital availability. Presumably, if they had not been able to secure their large initial capitalization, they would have targeted a smaller niche.

Compaq's manner of product introduction reminds me of the relentless Japanese style that will be discussed later. They quickly roll wave after wave of technology based products into niche markets. *Study, Target, Niche, boom, Niche, Boom, Niche, BOOM!* The process is effective.

Compaq's product quality is impeccable as well, and anyone who can ramp a company up that fast has to be excellent at manufacturing. Compaq is the only significant U.S. supplier of portable PCs, but recently Texas Instruments and some others, most notably Airis Computer, have entered the field with some very attractive products.

Comment: Zenith is a familiar U.S. brand of portable PCs, and Zenith's products enjoy widespread usage by the U.S. government. As they say, "The quality goes in before the name goes on." The statement is literally true since the products sold under this brand name are now Japanese, and Zenith Data Systems was recently purchased by France's Groupe Bull. Airis, incidentally, is a new venture formed by Zenith Data Systems alumni.

some equity. What if the unthinkable happens and your venture fails? You get a write off, experience, and valuable training.

Note: To be sure, it is not easy or everyone would be doing it. There are many troublesome details. Your staff is the first problem. They will scream to the heavens if you spend "their" money on a venture, since they want to use it for other things. Consider giving them some stake (a bonus?) in the venture's success, but do not contemplate running the venture by committee if you hope to solicit investment. Your odds of failure are very high unless your team includes senior people with high-tech venture experience, and such resource is in scarce supply. If you can find it, consider the use of outside talent (except the leader) to lower this risk. You will personally need to protect and nurture the venture initially. Selection of partners needs thorough consideration, as does the timely solicitation of funding. Unless you choose to do it all yourselves, you must have a good exit strategy for your investors.

What success rate should you expect from your ventures? Well, it depends. If *all* your replacement products don't succeed, something is wrong. Here, you are working in well known areas, so failure suggests a level of inability in your organization that should be dealt with harshly. Extensional and new product ideas have

"If you are under control, you are not driving fast enough (to win)."

— *Mario Andretti*
Race Driver

How little is too little?

We have an unusual assignment where we are working as a team with another consultant on a turnaround. The client had a troubled history of episodic management changes, where the innovative President was ousted and replaced by administrators backed by money. When the administrators failed, the President again took over. The company was at the end of the latter cycle. The President had been running it with woefully inadequate funds, and the company was in desperate shape and about to close its doors. Without the President's vision the company failed, but with him in charge it had never made any money. This is an easy client to like, but a hard one to help. The President is honest, creative and caring, and he inspires loyalty. Unfortunately, he is a better leader than he is a manager. He has problems with delegation, often ignores business details, and enjoys invention more than completion or competition. He has spent most of his energy on creating clever new things, not on implementation and production. Instead of replacing the President, we agreed to form a three person executive committee with him. The company's board approved this, and based on this agreement the stockholders put in enough money to buy a few weeks.

The President chooses the general direction, but we twist this toward profitable business areas. If he wants to go North we normally acquiesce, but the executive committee must approve all policy and financial decisions. The committee helps the President select the trails and avoid the swamps, but it is his company to run. This arrangement was deemed preposterous by knowledgeable outsiders, since conventional wisdom dictated replacing the management and bringing in more investment.

Within weeks we had helped move the company into the black, though it is a struggle to stay there. The company is pushing a large ball of past financial obligations ahead of it, and has an unfriendly banker. Such a business situation is difficult. (This may be a pro bono project, since we are taking our payment in stock.) The company is on a COD basis with its suppliers, and lacks the money it needs to buy materials so it can ship products. Volumes are low, so costs are high and margins low. Funds are drained by debt service, and a painful premium is paid because parts must be purchased in low volume and on an urgent basis. There is no money for promotion or design improvement. We are trying to break that cycle, but this is one of the tightest situations I have ever worked in. Our best hope is to provide funding from customers and distribution channels until we can get the client's accounts current and do some debt restructuring.[28]

For the purposes of this example, I want to focus on what can be done creatively with venture techniques even in a situation this desperate. In a month the client had been repositioned into two related, new, and more profitable business areas. By the end of the second month we showed prototypes of a new product, and by the third we were shipping. We have a new distribution channel for this product under contract, and we got the distribution company to fund its development.

A second, low volume, high margin, related product is possible. Again, the client lacks money to develop the product. We may have interested other parties in forming a new venture, a service company that would use this innovative product as its basis for competitive advantage. Because of the clients troubled financial situation, it may be easier to bring equity investment into a separate venture. The venture would be a customer for the original company, our client, but it would be separately funded and managed. A third possibility is that the client may be able to induce key customers to front the money for improving and extending existing products. This also is under investigation.

The outcome is still in doubt, but the situation is improving, and it will make a good story if we succeed. Our goal is to save the company, but to leave it in a form that the President can own and identify with.

---

[28] We are, of course, doing the normal financial management things, including close review of accounts receivable and payable. We have moved to net-net terms, and, in some cases, prepayment.

(much) higher degrees of risk and uncertainty, so the early phases are really somewhat inexpensive searches for knowledge. It's not wise to punish these early failures unless foolish or repeated mistakes are made. You should encourage risk taking, and early failures should be rewarded if valuable knowledge is gained.

Inexpensive early failures can be educational. The first company I started fizzled in its first two business areas, but the third one started an industry and made a good return. The total capital investment for all three was $35,000, and when the dust cleared the investors were happy. I would not want to repeat the experience, but admit it was a low cost

---

Is it practice, or is it real?

When working for Tektronix in the late 1970s, I had the responsibility for replacing a high volume product line. The product line was called the "400 series," the core line of portable oscilloscopes. The products were aging, but still producing several hundreds of millions of dollars in revenue annually. Two attempts to replace the line didn't get off the launching pad, management had been replaced, and distrust was extreme between marketing and engineering. All the key players were new, and we had not worked together before.

We did a simple product for practice, Tek's first catalog digital oscilloscope, called the 468. It used commercial parts, and was an addition to a standard product. The development cost was low, so the project was easily justified and quickly implemented. Somewhat to my astonishment, the product turned out to be one of Tek's "top 10" revenue producers. It remained so, and it was still growing in sales when it was phased out for production reasons. The major product line that followed was also successful.

Where possible, *it's safer to move into the future through evolution than by revolution*. I talk much about innovation and doing new things, but I also advocate continuous improvement. Incremental moves build the foundation of revenue and experience you will need to compete in the "suicide square."

---

way to train young talent. We came close to losing the venture. I don't suggest a 66% failure rate as acceptable. A seasoned team can get close to a 100% success rate, but you can't start there. I often pick simple projects for "warm up," to let the team work together and enjoy a success. The interesting thing is that you don't know how good the payoff can be until you do one of these small "practice projects."

Doesn't management process help?

A large, financially strong, industry leading, electronics company grew concerned that they were falling behind in their markets, and their products were aging. The Executive VP directed that the operating divisions would henceforth be measured by the percentage of their revenue that came from products less than three years old. Goals were set, and extra incentives given for "high impact products" that produced large revenue. The goals were soon met, then exceeded, and charts of "product newness" were compared annually for key managers and employees to see.

Five years later, the company was in much worse shape in every market, financial performance had deteriorated, and lay-offs had started. The divisions had replaced their old products with "new" old products. These were essentially the same products, but with minor improvements and repackaging. This was the safe decision, since in each case one could be assured of getting all the revenue from the old product plus more, and the markets were large. Unfortunately, competitors had created new markets based on new technology. The company found itself in a hopeless position, caught between those who sold the old types of products cheaply and others who sold unique products at premium prices. In marketing jargon, they had been positioned by their competition as the leaders in expensive old products. One reviewer made the following comment, based on this example. "If you paint horse manure gold, you don't have gold. You only have prettier horse manure."

Comment: Note the need for product and market understanding. Traditional "industrial age" management process had exactly the opposite impact of that intended.

What failure rate you allow between the concept and the prototype depends on the wealth of your organization, your culture, and the cost of the program(s) in question. Different companies have different objectives. Some want many small wins, while others pursue larger opportunities and accept higher failure rates. For example, Sun reportedly uses the latter strategy and Cypress Semiconductor the former, but both are growing at compounded rates of over 100%.

For an important market, it may be desirable to start several programs in parallel, and let the best one win.[29] Here you are making a trade off between cost and risk. You are increasing your development costs several-fold, but increasing your odds of winning by the same amount. Whatever your strategy, management tolerance should cease after the prototype is completed and beta phase testing conducted, and at this phase you should demand 100% success. The purpose of prototypes is to resolve technical and market uncertainty. The time to kill programs is at the prototype or sooner. Product failures after introduction are very expensive, and should be avoided.

---

[29] In World War II this was the process used for new aircraft design. Opposite design approaches were funded to prototype and tested against each other. Interestingly, often more than one approach worked out and was put into production. In fighter aircraft the P-51, the P-47, and the P-38 had radically divergent design philosophies, but all were successful. The process worked for bombers too. The B-17 was opposite to the B-24, and the B-25 to the B-26.

Comment: It's always interesting to me that most companies prefer experience, but abhor mistakes. The most valuable experience requires allowing mistakes. We practice this in school and when raising children, but we tend to be intolerant of mistakes in business.

High Tech ...

Role models

To exemplify the approach I am advocating, I searched for successful companies that practice these methods consistently. I selected two, and chose these because they are dissimilar in many ways, and almost philosophical opposites. One is an archetypical high profile, maximum growth, VC backed, Silicon Valley company, with a tough, charismatic, and controversial leader. They run a tight ship and favor formal systems and procedures. The other is a small, privately held, laid back, Oregon company. Their leader is participative, and they favor consensus and informality.

They have similarities too. Both do technology-enabled products, both are leaders in their markets, and both have near perfect records of new product success. Both leaders would, in fact, take great exception to my use of the words "near perfect." Each was absolutely firm in will and purpose that *every* product they want to do and approve must be successful. Each contends that their company's track record at implementing such products was not near perfect, it was perfect.

The low road

A friend, Robert Metzler, and three associates started an Oregon corporation called Audio Precision in 1984. They found a small niche, and sacrificed for several years. They *avoided* professional venture funding, because it wanted to back only "big wins." They financed their company with their own money through the prototype, and then floated debentures with a stock "kicker" and five year payback. Capitalization was thin, $150K of their money, and $350K of debt. The debt structure was creative — a savings and loan, a Small Business Investment Corporation (SBIC), and a private investor participated equally in a debt package. The debt was repaid in 2 1/2 years, and by 1990 they had a successful company with 23 employees and $5M in sales. The employees own 89%, and the company is highly profitable and can self fund its growth. The S&L has exited, presumably because it was required to by regulation, but the others still enjoy quarterly dividends.

As the company's name implies, their products measure audio phenomena *precisely*. It is a niche large enough to afford growth, but too small to interest multi-billion dollar companies. The market is well protected, and offers room for expansion. In their area of expertise, they have perhaps three or four of the world's leading half-dozen technical experts and they know their market and their key customers intimately. The founders want to make money, have fun, and avoid administrative tedium. They could quickly become a $50M business, but the plan is to limit size to one layer of management and about 45 employees.

The four founders have varied backgrounds, but share mutual trust and long experience in working together. Bob is technical, has a marketing background, and has long been close to this market area. Reportedly, the founders usually concur on major business and new product decisions. Bob serves as the President and CEO on paper, but the company is run informally. When interviewed for this book he said, "Essentially we all have veto power. If any of us is acutely uncomfortable with something, we don't do it. We generally agree, and there have been very few cases where any one of us felt so strongly about something to inflict his will on the others." They are informal, but they are not casual. It's their company, they care about it, and they are deadly serious about results. Whether I could mention the name of their company in this book was put to a vote of the partners, as are all significant decisions. They don't make a big deal about control and tracking performance, but they target and implement their products well.

I asked about success rate. The answer was emphatic, "We have successfully finished and introduced **every** product that we started." Depending on how one counts, this amounts to either one major product with many options and extensions, or seven separate products. Anyhow, at least seven separate engineering and product introduction cycles have passed successfully.

This company's planning and control process is loose. These are people who left a highly structured but ineffective Fortune 500 company in search of something better. Their vision of a better company is one that avoids bureaucracy, meetings, and process. To plan products, a large group — all the company's professionals — offers inputs, opinions, comments, and sometimes specialized expertise. There are no formal proposals, just discussions. The founders listen, review what has transpired, and decide what to do. When there is oral agreement, they write down the product description and its specifications. Then they do it. That's all there is to describe: *They decide, write down the specs, and then they do it.*

"What of budgets and schedules?" I asked. "We don't have any." "You don't have any budgets or schedules?" I asked, incredulous. Bob patiently explained that most of their R&D expense was their people's time. The company has the tools to track expenses by project, but they don't bother. They do budget and track their "outside" expenses such as promotion, trade shows, and such. They may start budgeting for their outside engineering expenses, and they are discussing this. Schedules are also a discussion point, and I sensed discomfort. The first products came out when expected, but the last two were late. This is felt to be because the company is getting larger, so the engineers were being diverted into more things. The impression I got was that the group agreed that this trend was disturbing and this problem needed solution. I sensed that they just thought there must be a better way to solve the problem than using (highly distasteful) schedules.

> **Note:** Why do these people resist the use of tools like budgets and schedules so strongly? Perhaps because they are so often misused by managers in large companies who substitute process for knowledge.

Since this was a good example (unusual, but good) of the methods advocated, I asked Bob if he would be willing to speak at seminars. He declined. "What is there to teach? It's basic. Just find a good niche market, get the best people, and do it." The fundamentals are simple, but few practice them. As you can see, Mr. Metzler is a private person who is low key, relaxed, and avoids publicity. His personal goal is to spend more time with his wife now that the company is doing well.

Observation: The small privately funded high-tech company is far more prevalent than commonly realized, and much of the job growth in the U.S. is coming from this. In Oregon, *half* the AEA member companies have fewer than 25 employees. Software and tiny companies are often not AEA members, so probably 70% of the technology-based businesses in Oregon are this small.

The high road

The premier Silicon Valley company as we enter the 1990s is Cypress Semiconductor. It's a VC's dream. Founded in 1983, Cypress has a compounded growth rate of 127.8%, the highest in the industry. It is now about $200M in size, produces $146,000 in sales per employee, and posts a 60% gross margin.[30]

Cypress' CEO is the outspoken T.J. Rodgers, and he does it by the numbers. His favorite movie is "Patton," his favorite book is Chuck Yeager's autobiography, and on his desk is a note pad proclaiming "From the desk of God."

Recall the "DRAM debacle." Rodgers almost single handedly turned political opinion against a government sponsored DRAM con-

---

[30] Most of this information comes from **Electronic Business**, April 30, 1990.

sortium, U.S. Memories, that he claimed would stifle entrepreneurship. This went against the wishes of the AEA and many large companies in the semiconductor industry, so he has critics.

Comment: I find it fascinating that Patton comes up so often as the role model for authoritative American leaders. Patton's style was unusual for at least three reasons.

- Contrary to his management's wishes, he invested to exploit success.

- He didn't always care who got the credit, if the job got done.

- He cared about his people, and used technology to get closer to them so he could help.

In one case when his superiors would not supply sufficient fuel for his Army to pursue a major breakthrough, he bought it with his personal fortune. After success was achieved, he billed them. Patton did not worry overly that his adversary had superior equipment. Instead he punched through into undefended rear areas and wreaked havoc with supply lines. He put so much pressure on his enemy that they had to come to him, but they rarely made it. When the enemy moved they were easy targets, and Patton worked closely with the tactical Air Force. He took great glee in tallying the tanks that never reached his forces.

I had the occasion during the Vietnam experience to be visiting the Army's Staff College in Kansas when General Eisenhower's funeral was held nearby. It prompted the old soldiers to reminisce, but to my surprise they talked of Patton, not Eisenhower. The theme was how Patton was always on the scene. He used radio and a small aircraft to stay close to his troops, and when he saw mistakes he would land next to his tanks and set matters right in his memorable way. Decades later, his commanders were still thankful that helicopters had yet to be invented.

Rodgers, like Patton, is authoritative, fastidious, smart, and charismatic. He is also extremely thrifty. "I will not tolerate people in this company spending our money as if it were something free." Cypress' employees fly coach class, and Rodgers himself drives a Honda Accord and buys a new one only every three years. He delights in curbing excess spending. Everything is controlled, and there has been a "turbo MBO system" since day one. All employees are measured against goals weekly.

Contrary to what has been reported in the press, Rodgers does not *personally* review each employee's performance weekly, and he states explicitly that he does not care to do so.[31] As his company grew, he organized it so that he doesn't need to anymore. "The goals are generated by the people on the bottom and

---

[31] Source: Private correspondence.

organized by projects on Mondays. On Tuesdays, the goals go through the system again where the supervisors organize the goals of the people working for them. In this manner, we get project-driven goals as well as functional organization controls." Anyone found wanting is warned and can be fired. There are rigorous systems for controlling head count, budgets, capital expenses, and product planning.[32]

Rodgers objects to reports in the press that he dictates that the company and all subsidiaries snap to attention and use his control systems. He explains that he does insist that all the subsidiaries start with and fully understand Cypress' systems, but he encourages continuous improvement, known as *Kaisen*. "Often times the changes of the subsidiaries are very good and implemented back into Cypress' systems." He does want his systems to be used and fully understood before creativity sets in, and he brooks no argument on that point.

Cypress' product planning process is highly formalized and uncompromising. A shaky product plan is not (quite) a firing offense as the press reports, but Rodgers' concedes that the project is "held hostage" until there is a good plan. He pointedly takes exception with trade press reports that "out of 160 new product plans approved since 1983, all but one of the products succeeded." He says instead that, "By ensuring that every product has a solid business plan signed by *all* vice presidents, we

---

[32] For a more detailed description, see "The Goal System That Drives Cypress," **Business Month**, July 1987, pp 30-32.

> "First rate people hire first rate people. Second rate people hire third rate people."
>
> — *Anon*

have achieved the record of **never** having canceled a project once we started it." Cypress is spending 25% on R&D, and they spend it well.

Their vision is to become a $1B company of smaller ($20M or so) niche businesses. Rodgers' calls the company a value-added VC, and the value he adds is control and guidance. "I will show you, guide you, so you *will* be a success," Rodgers has been quoted as saying. Again, he doesn't want this point misunderstood. He says he doesn't pretend to supply any personal "magic" to ensure new ventures will somehow automatically be successful. His point is, "We will not tolerate failure. The (quoted) sentence simply means I will do whatever is required to make each venture a success."

Almost everything you read about Cypress centers on Dr. Rodgers' personal leadership style and their superb business results. I wondered how they get their guidance and their intelligence. What forces keep them on course and in touch with their markets?

Comment: Back to Patton. His victories were attributed to his personal leadership. When asked how he knew what to do he told wild stories about reincarnation and visions, but the reality was that his intelligence information was superb. He had modern intelligence, he knew his adversary's traits, and sometimes he even had their codes and plans. He also used historical information effectively. His tank columns would sometimes bypass roads and unexpectedly appear in the heart of enemy territory without encountering resistance. His intelligence people collected ancient Roman maps, so he knew where the passages and river fords were before roads existed.

It's not well known, but Rodgers has a Marketing Guru of the first rank on his staff, who balances his quest for action with "outside in" market knowledge. Little is said about Lowell Turriff, Cypress' V.P. of Sales and Marketing.[33] He is reportedly formal, mannerly, understated, and analytical. He keeps his cool, speaks softly (at least compared to Rodgers), and knows whereof he speaks. While Rodgers wants quantified detail and action, Turriff wants to know what his customers think and feel. Turriff does his job well, and he makes sure his leader is close to his marketplace. When the VCs said they liked Rodgers' technical vision but wanted a more detailed business plan, he wrote one in a weekend. They got their funding, and Cypress was off and running.

When Cypress got technically arrogant and wandered from what its markets needed in 1986, Turriff presented Rodgers with four hours of back-to-back horror stories — from missed deadlines to lost accounts. The situation was serious. Disgruntled customers were starting to "design out" Cypress, and this is

---

[33] Source: CEO's journal, "The Boss's Best Friend Is a Mean Alter Ego," **Business Month**, May 1990, by Michael Rogers.

death for a semiconductor house. It must have been an interesting meeting. Rodgers listened, believed, and fixed the problems. "That's when we became a [market driven] manufacturing company, rather than an engineering house," Rodgers recalls. The two sometimes clash, but they make a world-class team. "T.J. and I probably look like we are fighting a lot," Mr. Turriff observes, "But underneath, we have mutual respect."

Cypress seeds start-ups just as the VCs do, and asks for a minimum of 80% ownership. Their model for the 1990s is a loosely knit amalgam of many small companies, all with access to a central pool of financial, manufacturing, and sales and marketing resources. When they needed design centers, they set up several and let them compete. Each center wins designs based on superior turnaround time and cost, and when a design methodology is proven to aid productivity, it's transferred to all the centers.

By reputation Rodgers is tough, but he's also capable and he cares. He inspires, he challenges, and the employees value his leadership. One said, "He could do every job here better than the person hired for the job." Others say he is always there when his help is needed. Is Rodgers *really* like Patton? There are obvious similarities, but Rodgers contends, "(writers) got carried away and made me appear a whole lot more macho than I really am."

Comment: Cypress Semiconductor is a singular case, but it has siblings with similar competency, intensity, and track records. Silicon Graphics,

Oracle Systems, and Sun Micro-systems are *all* growing at com-pounded rates of over 100%. As far as I can tell, the cultures and styles of these firms are every bit as indi-vidualistic and dissimilar as the examples I have already presented. Their styles differ and evolve, but they all practice the basics, and they all depend on having fast, knowledge-centered, independent teams.

There are mature, larger, U.S. elec-tronics firms that succeed consistent-ly too. Growing at around 15% com-pounded doesn't sound like much, but firms like HP do this, or better, with regularity. At HP's size of $12B, that represents $1.8B in new growth (nine companies the size of Cypress) every year. In the middle you have com-panies like Compaq that is, at a size of $3B, now slowing to a growth rate of about 50%. This is still $1.5B in growth each year.

High Tech …

Why we fail

The fundamentals are simple. Form a separate team, pick good people, empower them, fund them adequately and in a timely manner, bend the technology to match the market need, implement well, have a portfolio, avoid wild gambles, adapt to your environment. Then persist, get close to your customers, and practice *Kaisen*. If you do these things you can succeed almost every time, but few do, so failure is common. I think there are several main reasons.

The first is acrimonious internal **conflict**, where we worry more about organizational turf and charters than about customers and competition. Western society is largely based on adversarial relationships. In the courts one side wins, and the other losses. This is true in corporate politics too, and those who have the most power or who give the most persuasive presentations often carry the day. Small groups making intuitive decisions and working as a team on long-range unclear things find it hard to survive. It's too easy to denigrate creators, to criticize prototypes, or to steal from the future for today's needs. For small teams to survive, strong protective barriers are needed. Unfortunately, the barriers encumber implementation, so they eventually die anyway. Xerox's famous research center did much of the recent innovation in computers, but the company was unable to tap it for products. Even within groups, conflict is common, and, in high tech, marketing is often perceived as an inferior subculture, a place to put failed engineers and tired sales people. Some ego-driven technologists would sooner fail than have these strange people interfere with their creativity, and the odds of a marketing Guru coming from (or to) such an environment are slim. An academic basis for training high-tech marketing Gurus is missing, so industry lacks a clear model for competency.

Another reason is our **short-term focus** and **high cost of capital**. Consider again the traditional development process, as compared to the compressed time-to-market model, and recall Figure 3. Remember, one arrow is skinny and long, and it takes years to get to the market. Everyone advocates reduced time-to-market, and the "fat arrow" of parallel activity allows this but spends money faster. Total project cost is about the same (shown by the total area of the arrows) with either approach since about the same amount of work needs to be done, but the "skinny arrow" spends the money at a slower rate. In a cost cutting company, the short fat arrows are soon pared to long skinny ones, so schedules slip. Worse, functions like marketing and manufacturing are pared to junior people or removed entirely. The risk of mistakes and mis-targeting increases dramatically. This is done although everyone knows it is not in the company's best interest. The return afforded by the "fat arrow" is a great deal better, but sequential development allows a lower "burn rate." It makes the accountant's score cards look better, but is a poor way to do product development for fast changing niche markets. Most companies flog their project teams to go faster, but release funds much too slowly to allow it. VC-style total project funding would achieve better results. Most new ventures are unprofitable for the first four or five years, and the higher capital costs are, the more pain this can cause. That alone causes much grief for the U.S. Another drawback of our high cost of capital and short-term focus is that even when we win the battles, we may lose the war. Much as with the fable of the tortoise and the hare, even when we penetrate attractive markets we often succumb to the desire to stop and take profit.

A major reason is that we have more **friction in our system** — more symbiosis, but also more parasites — than those we compete with. In all Japan there are about 15,000 practicing lawyers, but here we have 606,000 and another 128,000 in law school and they all need to put bread on their tables. It's dangerous to be an officer or a board member of a public high-tech company, since some law firms make millions in contingency fees from class action stockholder suits. There are even "pet clients" that buy some of every traded stock, and they roll in and out of these stocks frequently. Top honors in this class go to a retired lawyer named Harry Lewis who has been party to between 300 and 400 suits in Delaware alone, but a Mr. William B. Weinburger specializes in high tech and has been a party to some 90 suits. If the stock goes down they sue saying "you didn't tell us it would go down," but there have been similar suits when the stock went up.[34] Generally the companies are insured, and stockholder litigation is set-

---

[34] "Getting mugged on the courthouse steps," **Upside** magazine, April 1990.

tled out of court for large sums. Private funding somewhat avoids this risk, but has others. For one, it's hard to raise enough money. For another, even non-participants can become casualties of the drug wars. If any investment money is allegedly drug related, your assets can be seized. Then there are torts, and, for instance, product liability insurance costs more than the manufacturing cost of a small aircraft. This is not yet a major issue for high tech, but it could be in the 90s as products target specific applications. Having insurance makes the likelihood of litigation higher. Even if you are insured or eventually deemed innocent, the issues are so serious that they consume your time and risk your personal fortune. Finally, there's government regulation, restriction, and bureaucracy. *Though we don't get all the government we pay for, we still get enough that it puts us at a global disadvantage in some areas.* Government and industry are often at cross purposes, and education, taxes, capital gains, trade policy, technology export laws, and government procurement regulations are all troublesome issues. Even setting up The Trudel Group was delayed because I was embroiled in major litigation as a witness. (It was against a foreign company for patent infringement, and my former employer prevailed.)

> **Note:** I don't mean to bash all lawyers. More than once attorneys have saved me from sharks, and I am grateful. Without courageous legal "shark fighters," small companies would not have much chance of survival. My point is that our system is much too litigious.

Another problem is our investor's **lack of intuitive market and technology understanding**. The U.S. doesn't treasure knowledge of the hard sciences, but you have to know intimately what you are investing in, and it takes more than resumes, a plan, and money to build a successful venture. Many of today's VCs have a tough time knowing how to invest in high tech, and most private investors find it totally confusing. Smart money is in scarce supply in the U.S. More than 50% of small high-tech U.S. company CEOs say capital is now their primary problem, while the large companies seem *afraid* to invest in new technologies or markets. In November of 1989, GE announced that they would spend up to $10 billion buying their stock back rather than taking a "wild swing" on an acquisition or investment in new technology. Wall street rewarded them for this shrewd decision, and their stock price went up. In similar manner, IBM has spent about $10 billion on stock repurchase instead of investing in laptops or supercomputers. Apple is still running by Steve Job's decade old product vision, but plans to repurchase 5 million shares of its stock. Honeywell, Tektronix, DEC, Chrysler, GM, and many others have invested major amounts in stock repurchase.[35] Long ago, David Sarnoff at RCA invested to create color television studio equipment, broadcasts, and receivers in defiance of his entire board. Today Sarnoff is dead, RCA is owned by GE, the Sarnoff research laboratories were sold, and the Japanese dominate television.

---

[35] For a good discussion of this, see "Are we eating our Seed Corn?" in the **New York Times** business section, May 13, 1990.

Our **quest for adventure** leads to failure in the world of high tech. Just as little old ladies lose their retirement money to crafty promoters or slot machines, so billion dollar companies, investment bankers, and VCs sometimes gamble on luck and seek adventure. The Japanese patiently invest in the future, but in the U.S. many would sooner gamble on the big win or the quick deal. The savings and loan industry crumbled because of rampant speculation, leaving the U.S. government — the taxpayers — faced with an

---

"Do you feel lucky?"

There's a famous line from one of Clint Eastwood's "Dirty Harry" movies. After a prolonged and bloody shoot out, the toughest of the bad guys has dropped his shotgun. The hero has subdued legions of foes, but his gun may be empty, and he may be helpless. The villain glances significantly at his fallen weapon. We all know what the scoundrel is thinking, but just in case there are some dimwits in the audience the scriptwriter has the hero explain it to us in detail. Dirty Harry then smiles, cocks his 44 magnum, aims at point blank range, and comments, "Before you do that, punk, ask yourself: Do you feel lucky?"

Do you really think that some dullards using management process or an inexperienced team of rookies will be lucky enough to win in the 90s? Do you seriously believe that such approaches might beat the likes of, say, Mr. Bechtolsheim and Sun, Mr. Rodgers and Cypress, or even Mr. Metzler and tiny Audio Precision with their smart, quick, dedicated, well-drilled teams? Most likely you will say, "Of course not. Perhaps we will become lucky and find a market with no competition. If we can't find one, we will beat up someone who is stupid and defenseless."

Well, I hate to disillusion you, but your presumptions are suspect. If you discover a worthwhile high-tech market without any potential competitors, give me a call and I will put you in my next book. (On second thought, please don't bother to call. Call a VC or a banker instead.) I must warn you that there are *many* other companies that have teams as good as the firms cited, and if you lack talent and knowledge you might accidental-

enormous multi-hundred-billion dollar (or more: some say losses are well over $500 billion) bailout.[36] Now we have a big mystery forming over how this terrible thing could happen, and who should be blamed. An investment firm, Neuberger & Berman, is running full page ads in **Wall Street Journal**.[37] The

---

[36] Source: **Wall Street Journal**, cover, July 3, 1990.

[37] Source: **Wall Street Journal**, July 25, 1990, page A13.

---

ly blunder into one of them. Compaq, HP, and the Japanese develop such teams consistently, companies like IBM develop them sporadically, and professionally backed new ventures would not even exist if they could not field such teams. The tough markets of the 80s were just practice for the 90s.

Much of what your Guru and your technical leader can do for you is to find and profitably exploit niches. You want to avoid the tough antagonists until you are established. Competitive positioning, in effect, means that you select the battle field so you have unfair advantage. Rational people should never run straight at a formidable opponent with a 44 magnum. Product positioning means that you get your *first* product close enough to customer needs to blast through FUD and competitive reaction. That way you have a revenue stream, and you can keep investing to stay ahead. That is as good as it gets, so let your competitors lose money while they try to catch up.

Some of the best niches can be found by "picking on" the large bureaucratic U.S. companies, but you should do this selectively, and choose your point of attack carefully. That method works, so find some market area a big firm has neglected, niche it, and invest to stay ahead. That's what Compaq did to IBM, what Cypress did to the big semiconductor companies, and what MIPS and Sun did to Motorola and Intel with their RISC chips. These winning companies had excellent teams and knew exactly what they were doing. The winners in the 90s will be the companies that niche well, move quickly, and keep going at speed. The cycle time for their next product will be less than for their first.

title is "Never Have So Few Taken So Much From So Many," and the subtitle cautions that "The speculators and their political friends ruined the S&L industry. Now, they have the power to ruin the stock market." The text explains that after the crash of 1929 and the ensuing depression, laws were passed to limit stock speculation with borrowed funds. This worked until 1982 when a new product was dreamed up, stock index futures. Because such futures are technically treated as commodities rather than stocks, they don't have the same "margin" requirements and are much cheaper to buy. Instead of investing, speculators can gamble on short-term trends with very little cash outlay. The advertisement claims that more money is now bet on the short-term direction of the market than is invested in the stocks themselves.

The last reason for failure is that **we have not adapted to our new environment**. Many of our largest companies still have **formal, rigid, outdated, bureaucratic control structures**. The information age demands fast, smart, flexible organizations. Most large companies have many levels of management that can say "no," but it's very hard to find anyone who can say "yes" until you work your way to the very top. In the environment of the 90s the situation may be radically different by the time you can get a decision. Even if sensible action results, there is no way to regain the lost time. The U.S. is still enamored of the Taylor-type management processes of the 30s. Mr. Taylor allowed us to build factories where a few people at the top think and make decisions, and the rest of the work force mindlessly implements according to detailed processes and guidelines. That may have worked in the mass-production based companies of the industrial age, but it doesn't work well in today's globally competitive, uncertain, and rapidly changing world. The shift to knowledge-based competition is revolution in the truest sense of the word, and the old business processes lack effectiveness in this new era. The greatest strategic resource of a company in the information age is not money or capital assets, it's empowered minds working from timely and correct information. Most U.S. corporations give lip service to these concepts, but few practice them.

In America many large organizations still prefer inexpensive and inexperienced, but docile and obedient workers who follow orders and pass information into the system for upper management action and decisions. Such a structure stifles innovation, but the theory is that it provides security for all. Someday you might get to be a manager and see the "big picture." You might even get a gold watch after 30 years of loyal service, unless, of course, you were laid off first because your employer could no longer compete.

Comment: The ultimate absurdity of Taylor-based process was Vietnam. Under Secretary McNamara, Washington decided where and when to fly, what targets to hit, and how many and what type of bombs to put under each wing. Sometimes they even announced the targets at press conferences. If anyone were so unkind as to shoot at you, there was even a policy manual, the rules of engagement (ROE), to consult before responding. The loss rates were monstrous, we lost the war, and I doubt Patton would have done it that way.

The innovators, your skunk works team and the early implementers, take high risks when they attempt creative innovation. Is it unreasonable to offer them high rewards if they succeed? Unfortunately, too many skunk works teams are expected to conform to corporate norms, procedures, and compensation structures.

There are probably other reasons, but in the end most of our failures are caused by **lack of knowledge**, **arrogance**, and **greed**. Success in high tech has always demanded technical knowledge. In today's markets this is not enough, and one *also* must view their products "outside in," through the eyes of the customer. What the customer wants or needs determines business success, and this may be very different from what the designers value. In the markets of the 90s technical arrogance will not lead customers to covet your products. Competitive companies in the 90s will combine technical, marketing, and manufacturing excellence. Excessive greed will put companies at a major competitive disadvantage in the 90s, so we must develop more financial "staying power" and sophistication. A successful technology-based new venture or product grows very rapidly, but it can easily take five or ten years to become profitable. *As markets become global those with vision and patience will have increasing advantage over those chasing quick returns.*

I think that by the end of the 90s our industrial world will look much different from what it does today. Many new companies will emerge, and even those whose names we recognize will look different in structure and culture. I think there will be major changes in the United States' financial and tax policies, and in our legal system, but I won't hazard a guess about how we will solve our problems. What will the typical technology company look like? This is hard to predict, and the best experts are having trouble sorting the winners from the losers. The decline of the large companies can't continue indefinitely, so something has to change. Some of the Fortune 500 will adapt and change radically, some will be bought, and some will not survive the 90s in any recognizable form.

I'm placing my personal bets on fast, flexible, knowledge-centered companies that value their people and let them grow, think, and create. Hewlett Packard and Cypress Semiconductor are good examples of firms that retain these attributes. Consulting firms like TTG will be there too, because we are creatures of the information age. In the 90s a small company often has more flexibility and greater access to knowledge than a large manufacturing firm. Partnerships will be common. The tiny companies may benefit from the clout and money of the "big guys," but the large companies won't prosper without exploiting the knowledge, skills, techniques, and experience of the smaller entrepreneurs. The U.S. is still strong enough in technology that we can be a leader if we make the right decisions. We have some tough choices to make, and our standard of living and position in the world is at stake.

High Tech ...

The future

High-tech managers perennially seek the "next wave." They ask, "What comes after workstations? What comes after PCs?" The future is not what it used to be, and the very nature of "wave action" will be different in the markets of the 90s.

There are major waves, but they are megawaves with names like ISDN, HDTV, and flat-panel color displays. The problem is that in today's global markets these are obvious and equally accessible to all, and U.S. companies are not well positioned to

> "We create our own future."
>
> — *Tom Peters*

ride the big surf.[38] Commerce Department reports on the ability of the U.S. to compete in the 90s are described as "bleak." Between 1983 and 1987 the sales of U.S. electronics companies electronics grew at 1/3 the rate of Japanese companies. Of the top five companies for U.S. electronics patent filings in 1987, three were Japanese, and one was European. A General Accounting Office report has identified that key portions of our industry have already been "hollowed out" by foreign acquisitions. The report (GAO/NSIAD-90-94) says things like, "Foreign investors may have bought U.S. firms not only to enhance their own position and acquire technology, but also to deny their U.S. competitors access to this technology." The Commerce Department

---

[38] Data from **Wall Street Journal** and **Electronic News** articles, June 11, 1990.

expects the Japanese to surpass the U.S. in the production of electronics goods by the early 1990s. Congress is holding hearings on these topics, but in the past government action has been neither timely nor effective. Everyone agrees that there is a major threat, and that if we lose the lead in high tech, we will lose the crown jewels of our industrial infrastructure.

Note: This could well become a major political issue of the 90s. The Harvard political economist Robert Reich has been doing studies. He asks Americans, "Which of these two futures would you prefer?"

a) Between now and 2000, the U.S. economy grows 25%, but the Japanese economy grows a whopping 75%.

b) Between now and 2000, the U.S. economy grows only 10%, but the Japanese economy grows an anemic 10.3%.

Except for one, the majority of every U.S. group surveyed has chosen (b). The exception was the economists, who voted for (a). Choice (b), of course, implies closing the borders to trade and investment from Japan.[39] We should think about what this data portends for the 90s. Intelligent people have become so desperate that they are willing to harm themselves seriously if they can do more damage to those threatening them.

---

[39] **Wall Street Journal**, June 18, 1990.

> "The government's protection programs are interesting. Wolves are a federally protected 'endangered species' in states that have no Wolves, but they are legally hunted and killed in all the states that do have Wolves."
>
> — *Comment made by tour guide ,Wolf Haven Tenino, Washington July 1990*

This is a tough problem for Congress. No one wants the U.S. to lose its best employment sector, but the issue is what to do to solve the problem. The large established companies lobby for government intervention, for tariffs, for subsidies, for consortiums, for cartels. They say that if this is not done with urgency, they may become as extinct as the dinosaurs. They say save the large companies, because that is where most of the jobs (and votes) are. They show data that in most large markets, the top ten companies usually control 60-80% of the market. Some even claim that small start-ups are draining them of the talent they need to be competitive. The smaller companies, horrified, point out that there is no evidence that direct government intervention has ever helped industrial competitiveness (they call it the "Amtrak solution"), and much that it can do harm.[40] Specifically, they fear that this would be using tax money to allow the big companies to compete with *them*, not with foreign firms. They say, "Hold on! Just because the big guys chose to under-invest in new things to make better short-term profits, why should they be subsidized to catch up?" They note that they are the *only* group that is competing effectively, and that most job

---

40 "The Congressional Hearings on U.S. Memories," WESCON Keynote Speech, San Francisco, CA, November 14, 1989, T.J. Rodgers.

growth comes from small companies.[41] They also point out that the major exporter to the U.S. is Canada (mostly logs), and the major investor in the U.S. is Britain (everything, but mostly not high tech), so why worry about Japan and not these? As usual, both sides have convincing arguments and eloquent spokesmen.

Perhaps both groups are right. As a consultant I would like that, because I value them both as clients and want them both to prosper. As an American, I think we need them *both* to win. There are some major issues that both sides agree on, and these might form the basis for helpful government action. These issues include things like long-term R&D tax credits, a (much) lower capital gains tax, and better education.[42] Education is a particularly troublesome problem, and in the late 1980s American high school students consistently ranked as the most poorly educated in all the major industrial countries in mathematics, language skills, geography, and all other knowledge measurements.[43] Despite doubling spending per student between 1959 and 1989, Scholastic Aptitude Test (SAT) scores dropped by 10% for verbal ability and 5% for mathematics over the same period. Using Los Angeles as a sample, violent crime

---

[41] According to **Business Week**, over 52% of U.S. exports come from companies with fewer than 100 employees.

[42] Opponents say tax credits were tried and didn't help. R&D tax credits have been on-again off-again so many times since 1981 that few trust the government to leave them in place.

[43] Source: **Business Month**, May 1990.

in high schools increased from almost nil in 1959, to an average of 15 crimes per day in the local school district in 1989.

I predict these issues will be attended to in the 90s, but I don't know if this will happen soon enough to help U.S. high-tech competitiveness. I predict that our global economic competitiveness will be a major issue — probably *the* major issue — in the 1992 elections.

I think we need better policies to encourage private sector technology investments and strategic partnerships. Can you imagine the wealth we could create if the U.S. could somehow combine the manufacturing muscle of, say, a Motorola or a Xerox with, say, the creativity and innovation of Steve Jobs' new company NeXT? It's too late now for a U.S. partner, since Canon just invested 100 million dollars in NeXT. (It's easier for U.S. innovators to partner with foreign financial and manufacturing strength.) Even without government encouragement, I think most large companies will be making more use of consultants and strategic partnering to increase their ability to innovate in the 90s. An interesting fact, one that has not received much mention, is that there is a clear pattern that seems to predict which countries will win at international competition. *George Gilder has pointed out that in every industry where we have more competitors and start-up companies we beat the Japanese, and vice versa.* The best example of this is the software industry.

Japan and Europe have been investing large sums for years to get ready for the next big waves. The Japanese say the U.S. "abandoned its factories to play the stock market." This is an overstatement, but clearly our technical arrogance of the 70s and paper entrepreneurship of the 80s won't win in the global markets of the 90s. The U.S. has lost share in every electronics market except software

> "I don't believe in entering manhood contests with the Japanese. The Japanese are like Sumo wrestlers — they're massive and they train diligently. If you run at them directly, you'll lose."
>
> — *T.J. Rodgers*
> *President and CEO*
> *Cypress Semiconductor Corp.*

and medical equipment since the mid-80s, and our decline persists. Something needs to change if we are to remain competitive, and this book is my attempt to effect change. The talent can be made available to rejuvenate the U.S. high-tech industry, and the right tax policies and better access to more patient capital might allow this to happen. If it does not happen soon, our standard of living will suffer.

I can't get too excited over the mega-markets, and I think the "wave model" of products itself is outdated and becoming suspect. I think there will be a multitude of opportunities, and I think the best will be niche markets. If so, the fundamentals described in this book will help. I think the winners will be quick companies that can combine excellent technology and marketing, and their products will be tightly focused on applications. Instead of talking about waves of product types, I think in the future we will be talking about targeting specific application areas. Where in the past only some 60% of high-tech products failed due to poor targeting, in the future market fit will be even more important. This is already true in Japan.

As we move into the last decade of this century, I think any player can win. In 1987 Japan exported $38 billion of electronics, the U.S. $35 billion, and Europe $26 billion. We are declining, and they are growing. The U.S. is behind in marketing, but ahead in sales and (so far) technology. The main U.S. strength may be innovation. We do new ventures better than anyone else in the world, but our tax laws and our cost and availability of capital are slowing this. The Japanese lead in market targeting and financial strength, and they have better manufacturing, more consistent quality, and the best global focus.

Not enough has been said about Europe. The twelve countries of the European Economic Community (EEC) were not major players in the technology markets of the 70s and 80s, but they have major strengths and are consolidating. The Europeans are getting ready for the global contest and doing things that few thought possible. The atmosphere in Europe is already dynamic, they call it "Euro-phoria." Americans think of Europe as the past, the old world, but it's time to adjust this vision. Europe is already a technology leader in the following areas: communications, transport aircraft, power generation, rapid rail transport, and television. They are tough competitors in TV, have the first and fourth largest manufacturers of TV sets, and are well positioned to pursue HDTV. They have the only supersonic and the only fly-by-wire transport aircraft, and excellent satellite and rocket technology. They are ahead in implementing digital communications, both ISDN and digital cellular radio. For short distances, rapid trains (300 kilometers per hour

"Why is everybody always picking on me?"

As we enter the 90s, the giant Japanese cartels (Toshiba, Hitachi, Mitsubishi, and NEC) lead the world in DRAM production. It's not as comfortable a position as one might think.

In the U.S. tiny Micron and larger Texas Instruments (TI) are doing well in DRAMs. TI was silent through all the Congressional hassles on the topic, because they were in an awkward position. Some DRAMs flooding in from Japan were from a TI plant there. They also have plants in Texas, and, now, Europe. There is much sand in Texas, and TI is happily converting it to chips. They are probably the largest non-Japanese supplier of DRAMs. As mentioned, the Europeans are making major commitments to leadership in DRAMs. On the other flank, Goldstar and Hyundai will each open plants in 1990 with a capacity of five million 1 Mbit DRAMs per month, and Samsung is already building six million a month.

The amusing thing is that the cartels are under fierce attack from a tiny Japanese skunk works called NMB Semiconductor (NMBS). It was started in 1986 by a leader named Takumi Tamura, who was young in mind and had revolutionary ideas. NMBS is now about the size of Cypress in sales, $227M, but has only 440 employees. Mr. Tamura was a troubleshooter for Sanyo, where he fixed manufacturing problems for their radios and TVs. They finally made him retire, and he got interested in electronic components. He thought he could build totally automated plants to manufacture DRAMs. The experts said it was not possible, and no semiconductor company would back him.

Undeterred, Mr. Tamura went to a friend, Mr. Takami Takahashi, who was President of an old line company called Nippon Miniature Bearings (NMB) that made precision bearings, motors, and keyboards. NMB helped Mr. Tamura get the bearings Sanyo needed for VCRs in 1981. Mr. Takahashi put up the money. Mr. Tamura didn't even have a DRAM design, so he bought rights to a DRAM designed by a Denver company called Inmos and worked with the designers to make it producible in high volume. Mr. Tamura had never built any DRAMS, but he was eager. At the age of 67 he flew over an undeveloped rural seaside resort called Tateyama in a helicopter, donned rubber boots and a construction hat, and started building a plant. Some $280 Million later, his new plant was producing hundreds of thousands of 256K DRAMs per month. The world shook.

Conventional wisdom said that it should take several years to build such a plant and over 1,000 workers to produce that volume. Mr. Tamura had only 48 workers, and his first plant was on line in six months. "Nobody, especially the Americans and the Europeans, can believe we did it in just six months," says Tamura. "What Magic did you use?" they ask.

The cartels were interested, but not worried. Mr. Tamura didn't even have a sales force. Why would anyone buy an unproven design from an unheard of company, when they could get all the DRAMs they needed from them? They could bury Mr. Tamura's company with their production capacity. Mr. Tamura found another friend, Andy, who needed some DRAMs. Andy Grove is CEO of Intel, and he may be a bit miffed at the cartels for dumping DRAMs and forcing Intel out of the business that they created. So Intel ordered some — all that Mr. Tamura could build. Andy signed a five-year contract making NMBS Intel's exclusive memory supplier and agreeing to take at least 70% of the DRAMs they could build. Intel does have a sales force, and they now have exclusive marketing rights to NMBS' DRAMs. It turns out that the Japanese cartels can't yet compete with Mr. Tamura because his designs are better than theirs.

Mr. Tamura has found a niche. Tiny NMBS makes the fastest DRAMs in the world and now owns 90% of the premium performance market. They have only two small plants, but are building three million 256 Kbit DRAMs and two million 1 Mbit DRAMs per month. A larger plant will be running in August, and this will produce much higher performance 4 Mbit DRAMs at the same rate. In 1990 they will be making $342,256 in sales per employee, and they now have an $85M cash surplus.

How can Mr. Tamura come up with better designs than the large cartels? The United States still has the world's best DRAM designers though we can't fund the $250 to $500 million needed for modern semiconductor plants ourselves. A bevy of small U.S. companies (Ramtron, Vitelic, Alliance Semiconductor, and probably others) are selling Mr. Tamura their designs, and so is Inmos, though they are now owned by the French. Mr. Tamura now has the best designs, and his manufacturing is the best in the world and getting better.

The cartels are hard on his heels, and they say NMBS is much too small (only 0.3% of the world's semiconductors) to survive. Toshiba, Hitachi, and Mitsubishi have their fast DRAM designs almost ready, and that will surely be the end of Mr. Tamura. The Japanese bankers agree, saying that memory "is a dangerous business with wide swings" and not a good place

(cont'd)

Why is everybody ... (cont'd)

for a new venture. Not only that, but NMBS may be a bad investment risk since Mr. Tamura is now 72. "What will happen when the charismatic Tamura leaves?" they ask. What does Mr. Tamura say? He has set a goal for NMBS to be the world's leading manufacturer of DRAMs by 1995. The cartels say that tiny NMBS can never get the capital needed to do this. Even though Mr. Takahashi's company NMB has extra cash (a billion dollars), he would never dare to risk it in DRAMs. Mr. Takahashi smiles and says he thinks his friend's chips look much like ball bearings, and he likes ball bearings. Tamura's financial vice president, Mr. Shinoda, comments that since going public last year the stock has risen 230%, so NMBS needs no money in the near term.

Perhaps not, say the cartels, but how will you get the money to build a plant in America to get past their "fair market barrier" and sell to Intel after we are filling the world with cheap chips? Andy has not promised to build a plant for you, has he? Well Andy isn't saying, but sometimes at night when it's very quiet I hear faint chortles drifting up from Silicon Valley. It might just be the wind blowing over the mountains. But then ...

Note: Some speculate that the teaming between Intel and NMBS is in jeopardy. Intel wants all the output of the new 4 Mbit plant, but NMBS' Japanese customers are vehemently protesting. They fear that Intel would give preference to its own microprocessor customers and leave them short.

---

or so) can move executives faster and more comfortably than air transportation, and the Europeans have the best in the world. They are behind in computers and microelectronics, and that worries them. They are fixing this, and have firms that rank 10th, 12th, and 15th on the global list of IC manufacturers.

Comment: We tend to think that the Europeans use public investment more than we do to fund their new technologies, but this is simply not true. The U.S. funds R&D almost equally from the government and industrial sectors, but West Germany's industry invests almost twice what their government does in non-military R&D. Japan's industry invests almost four times what their government does in nonmilitary R&D.[44] Naturally these countries have government policies to encourage such investment, but that's a separate matter.

The EEC members view electronics as key to their economic survival and, perhaps, their national sovereignty. They watch what has happened in the U.S., and they play hardball. The Japanese tried dumping DRAMs into Europe, but negotiated terms and raised prices after being slapped with a 60% tariff at the EC frontier. *Interestingly, the tariff approach didn't solve the problem.* Despite the higher prices, Europe desperately needed IC memory chips for its industry. Japan had the needed chips and tripled its European market share from 1983 to 87.[45] There are now several Japanese IC fabrication plants in Europe, so overcapacity exists and a severe price decline of DRAMs is expected. In the end they will fight it out in the trenches, and the winners will be globally competitive on technology, price, and quality. Europe wants leadership in ICs, and they are investing large sums. Siemens and IBM are now working jointly to develop 64 Mbit DRAMs, and Siemens is a more formidable predator than little Micron Technology.

---

[44] Source: OECD, National Science Foundation, DRI/McGraw Hill, Federal Reserve Bank of New York. Quoted in **Business Week**, June 15, 1990.

[45] Source: **Electronic Business**, June 25, 1990.

This is a major five-year research program with costs split equally, and when it's finished, each firm will be free to produce these leading products. The Europeans are eager to work with U.S.

> **"The basic problem in America today is a lack of vision of the future."**
>
> — *Dr. Robert Noyce*
> *Inventor, Entrepreneur, and*
> *Statesman*
> *Quoted in* ***Upside****, July 1990*

companies, and I predict this will be our largest electronics market by the end of the decade.

It's interesting that some think that new economic theories need to be developed to model high-tech markets. As early as the 1930s, Oxford's Nobel laureate, John R. Hicks, speculated that there might be some product-market areas that could lead to limitlessly increasing economic returns. This theory has come up several times since, but the researchers abandoned investigations because these effects could only be modeled with very complex equations that were almost impossible to solve without super computers. Modern knowledge-based products are unusual, and some think this could lead to economic perpetual motion. Demand for these is almost unlimited if the products are cheap enough, and though semiconductors and computers require large investments, the per unit costs become small *and keep decreasing* if the production volumes are high. Today's European and Japanese government policies assume that high tech is somehow unique and worthy of special attention, but the U.S. government does not yet accept this theory.[46]

Both Europe and Japan are investing heavily in U.S. technology, both are spending more

on non-military R&D than we are as a percentage of GNP, and their rate of increase in
R&D investments exceeds ours. Everyone is,
of course, copying Japanese quality and manufacturing processes. Today, some of the
smallest companies have the best technology,
because they can move quickly and effectively.
With modern technology you no longer need to
mass produce items to build them cheaply. For
most products the economic-build quantities
are small, so you can afford to provide each
customer with exactly what they need and
want. Product differentiation and rapid global
distribution will be keys to winning, and production and quality will be keys to holding
market position. Products will be adapted to
local markets and to specific industry requirements. Companies will become smaller, flat-
ter, and faster, and the key assets will become
knowledge and enabled minds. Few of today's
market leaders will lead the markets of the
90s, and teaming between small and large
companies will occur across all geographic and
national boundaries.

All this has been much discussed, and
many are describing these changes as a transition from the "industrial age" to the "information age." Some good books to read on the
subject are Gilder's **Microcosm**, and McKenna's **Who's Afraid of Big Blue**, but the phenomenon was predicted years ago in **Future
Shock** and **Megatrends**. See the bibliography for additional recommended read-
ings.

---

46 **Business Week**, June 15, 1990.

High Tech ...

Some still doubt, but the information age exists, it's here, and it shattered the Berlin Wall. World governments are now interdependent, so trade barriers, borders, and Iron Curtains increasingly obstruct their creators. In the 90s world power and prestige will be measured by global competitiveness, not by military force. I think that's hopeful. These are exciting times, and technology already allows a tiny company tucked in the Oregon mountains to do high-tech business consulting on a global scale.

Comment: TTG was retained by a *Japanese* client to write an article so U.S. companies could learn how to better export to Japan and thus help reduce trade friction.

Acknowledgments

Some who offered support or education were Kevin Considine, Tom Dagostino, Phil Crosby, John Stoltz, Lew Terwilliger, and my classmates in Tek's now extinct Manager of Managers program. TTG's clients deserve special thanks, for they allow me to practice my craft and continue my learning. I must mention Howard Vollum and Masanobu Tada, two visionary leaders of the 60s and 70s that I was privileged to know and to learn from. Especially I must thank Langford Metzger who put up the seed money for the first company I started.

I want to thank my colleagues around the industry who selflessly took their valuable time to offer criticism and review drafts of this book, and to Professor Bob Davis of Stanford's Graduate Business School for mentoring and writing the foreword. I also give special thanks to A. Dale Aufrecht for technical editing and formatting, and Jim McGill for the illustrations.

Some of my best learnings came from watching honorable competitors — especially HP and the Japanese — and from people who touched my life briefly in classes, seminars, and as customers in a variety of jobs. I want to thank these people, and apologize to those not mentioned explicitly. Naturally, the author alone is responsible for all opinion, errors, and omissions.

High Tech ...

Selected Glossary

Arrow, Fat

The concept of managing product-development programs in parallel. All functions contribute and have roles and responsibilities throughout the project.

Arrow, Skinny

The traditional, engineering-centered, sequential, product-development process. Engineering conceives and develops, and the other organizations become involved later to manufacture and promote the product.

Beta testing

The concept of selling early pre-production units to key customers to solicit feedback, criticism, and testimonials. A variant is alpha testing where customers are loaned prototypes for early evaluation. This testing is usually done selectively and under non-disclosure agreement.

CAE

Computer Aided Engineering. Special computer-based tools to help engineers design, test, and evaluate products. This is a major industry, and the word is usually used to denote tools to assist electronics engineers. Sub-specialties include Computer Aided Software Engineering (CASE), Computer Aided Mechanical Engineering (CAME), and others. A related concept is Computer Integrated Manufacturing (CIM), where the desire is to use computer technology to move designs directly into manufacturing.

DRAM

Dynamic Random Access Memory, today's most widely used type of computer memory, pronounced dee-ram. Intel pioneered this market, and their first mass produced product was the 1103, a DRAM. The larger companies invested to perfect more traditional types of memory, but Intel ventured into DRAMs. The DRAM is a kludge that barely works, but turns out to be brilliant because it consumes little power and can be easily mass produced at incredibly low cost. DRAMs leak badly, and so can only remember their memory states for the briefest periods of time. Like a leaky ship that never sinks, today's computers must continually refresh their memories before they forget. The D means that this type of memory must be Dynamically refreshed at clock intervals measured in millionths of a second. There are many other types of computer memory, including Static RAM (SRAM), Read Only Memory (ROM) of various types, and non-volatile but much slower mass-storage memory such as the various magnetic disk drives and optical disks, but DRAMs serve as the working memory of today's computers.

FUD

The classic marketing strategy used by IBM based on Fear, Uncertainty, and Doubt. It serves to delay acceptance of new technologies and products until the larger, slower, companies can get their versions ready. Sometimes used as a verb, e.g., "We FUDed them."

Guru

Used in this book to denote the high-tech marketing equivalent of a Chief Scientist or Chief Engineer. A senior, seasoned, technically competent and business knowledgeable expert. The Guru's job is understanding market needs and working with technologists to fill them profitably and in a way that yields competitive advantage. Sometimes this results in market creation. There is no model for such expertise in academia.

Kaisen

A Japanese word and concept, signifying a process of continuous incremental improvement and an unending search for perfection.

Kludge

A system or product made up of poorly matched components. A "Rube Goldberg device" that is functional but bizarre.

LANs

Local Area Networks, a new form of communication that developed outside the purview of government regulation and created a large market. LANs are based on the concept that if one computer is good, many might be better. LANs allow people who work in groups to share information, software, and computer resources. LANs come in many different brands and types, and are used to link PCs,

MACs, and workstations. Many small computers, when linked together properly, can empower users to surpass the processing ability of large, expensive, mainframes and mini-computers. Conversely, LANs could lead to electronic sweat shops where managers count each keystroke of each employee. Whether LANs will further empower personal computing or return computing to the control of central authority is an area of debate and conflict.

MAC

The Macintosh computer. Steve Job's famous product, sold by Apple Computer, based on Xerox's Palo Alto Research Center technology, and implemented using Motorola's microprocessor architectures. The product is touted for good graphics and ease of use. MACs, like PCs, come in many versions of varying power and capability and are software compatible with each other. The MAC enjoys a solid niche (about 10% of the market) and has many loyal supporters.

PC

The ubiquitous Personal Computer. Generally denotes the versions spawned by IBM based on Intel's microprocessor architectures and Microsoft's Disk Operating System (DOS). These are the industry standard, and compatible products are available from many vendors. The early versions are to computers what the Model T was to automobiles, but the PC is now evolving into very powerful and modern machines. Modern versions offer high performance, large memory, networking, and excellent graphics; but they are still software compatible with the older and more basic machines.

PCjr

Not to be confused with the PC. A different, non-software compatible, lower cost personal computer developed by IBM for the home use market. The product failed dismally.

Shiva

The internal code name used for a specialized product in one of the examples. It was so named because the device achieved high performance by using a non-Von Neumann or parallel computing architecture. The idea of having many computers working independently reminded one of the project team (A Vietnamese) of the Hindu God with many arms. Whether we selected the correct God or spelled the name correctly are conjectural but unimportant.

UNIX

A standard computer operating system developed by AT&T. It has the advantage of being very portable, meaning that it can be moved to (hosted on, ported to) many different types of computers. There are several competing operating systems, and which will win in the 90s is the subject of almost religious debate. A computer language called C goes hand-in-hand with UNIX, and is also very popular in some circles.

Workstation

A powerful single-user computer used by professionals to do their jobs. In the late 80s Sun led this market, and their products run UNIX and use special "reduced instruction set" (RISC) microprocessors to achieve high performance. The distinction between the powerful PCs and MACs and "true workstations" is a topic of heated industry debate. Workstations have exceptional performance, excellent graphics, and networked communications capabilities, but so do some PCs.

Selected bibliography

Botkin, James; Dimancescu, James; and Stata, Ray: **The Innovators**, Rediscovering America's Creative Energy, Harper and Row, New York, 1984.

This is not particularly high tech, but it provides interesting insights about innovation.

Davidow, William H.: **Marketing High Technology**, An Insider's View, The Free Press, A division of MacMillan, Inc., New York, 1986.

One of the two fundamental works on high-tech marketing.

Deal, Terrence E. and Kennedy, Allen A.: **Corporate Cultures**, The Rites and Rituals of Corporate Life, Addison-Wesley Publishing Company, Inc., Reading Massachusetts, 1982.

When the outside world is chaotic and you cannot manage every detail, a healthy culture can provide guidance.

Drucker, Peter Ferdinand: **Innovation and Entrepreneurship**, Practice and Principals, Perennial Library, Harper and Row, New York 1986. (Also available in hardcover from Harper and Row.)

Drucker is the master of business process, and this is must reading for would-be entrepreneurs.

Gilder, George: **Microcosm**, The Quantum Revolution In Economics and Technology, Simon and Schuster, New York, 1989.

This is a good treatise on how the basis of competition in technology-enabled business has changed.

Hayes, Robert H. and Wheelright, Steven C.: **Restoring our Competitive Edge**, Competing Through Manufacturing, John Wiley & Sons, New York, 1984.

This provides a good start on modern manufacturing science and philosophy.

Jacobson, Gary and Hillkirk, John: **XEROX, American Samurai,** MacMillan Publishing Company, New York, 1986.

How Xerox learned through painful experience what Dantotsu (striving to be the best of the best) means, and how they can now successfully compete toe-to-toe with the Japanese at their own game.

Kami, Michael J.: **Trigger Points**, How to Make Decisions Three Times Faster, Innovate Smarter, and Beat Your Competition by Ten Percent (It Ain't Easy!), McGraw-Hill Book Company, New York, 1988.

Mike has done it all, from strategic planning at IBM to several entrepreneurial new ventures and turnarounds. He makes money and has fun. You probably can't afford his consulting rates, but his treatment of gap analysis alone makes this book worth reading.

Kouzes, James M., and Posner, Barry Z.: **The Leadership Challenge**, How to Get Extraordinary Things Done in Organizations, Jossey-Bass Publishers, San Francisco, 1987.

Success and leadership seem inseparable, but much of what purports to be written about leadership is really about management process. This book is about the practices leaders use to turn challenging opportunities into remarkable successes.

McKenna, Regis: **The Regis Touch**, Million-Dollar Advice from America's Top Marketing Consultant. Addison-Wesley Publishing Company, Inc., Reading Massachusetts, 1985.

The other fundamental work about high-tech marketing.

McKenna, Regis: **Who's Afraid of Big Blue**, How Companies are Challenging IBM — and Winning. Addison-Wesley Publishing Company, Inc., Reading, Massachusetts, 1988.

This describes how to play in niche markets against tough competitors.

Peters, Tom: **Thriving on Chaos**, Handbook for a Management Revolution, Alfred A. Knopf, New York, 1988.

This is a fundamental work about loving your customers and how to practice Kaisen in the United States. Tom Peters is as disdainful as Professor Drucker about high tech. He thinks uncertainty is an excuse, but give him a running business and he will show you how to improve things.

Porter, Michael E.: **Competitive Strategy**, Techniques for Analyzing Industries and Competitors, The Free Press, A division of MacMillan, Inc., New York, 1980.

A good book to provide top-level insights about where you might best choose to compete.

Suzaki, Kiyoshi: **The New Manufacturing Challenge**, Techniques for Continuous Improvement, The Free Press, A division of MacMillan, Inc., New York, 1987.

A Japanese tries to teach Americans the Kaisen that he learned at Toyota and Toshiba.

Selected index